ASYMPTOTIC DIFFRACTION THEORY AND NUCLEAR SCATTERING

Scattering theory provides a framework for understanding the scattering of waves and particles. This book presents a simple physical picture of diffractive nuclear scattering in terms of semiclassical trajectories, illustrated throughout with examples and case studies. Trajectories in a complex impact parameter plane are discussed, and the importance of the analytic properties of the phase shift function in this complex impact plane in the asymptotic limit is stressed. Several new rainbow phenomena are also discussed and illustrated. Written by Nobel Prize winner Roy J. Glauber and Per Osland, an expert in the field of particle physics, it illustrates the transition from quantum to classical scattering, and provides a valuable resource for researchers using scattering theory in nuclear, particle, atomic, and molecular physics.

ROY J. GLAUBER was Mallinckrodt Professor of Physics (Emeritus) at Harvard University and Adjunct Professor of Optical Sciences at the University of Arizona. In 2005 he received half the Nobel Prize in Physics for his contribution to the quantum theory of optical coherence. He has received many other awards, including the Albert A. Michelson Medal from the Franklin Institute (1985), the Max Born Award from the Optical Society of America (1985), and the Dannie Heinemann Prize for Mathematical Physics from the American Physical Society (1966). He was elected a Foreign Member of the Royal Society (1997) and an Honorary Member of the Optical Society of America (2008), and served on the Advisory Board of the Max Planck Institute of Light.

PER OSLAND is Professor Emeritus at the Department of Physics and Technology, University of Bergen, Norway. He is a member of the Royal Norwegian Society of Sciences and Letters and a Fellow of the Swedish Collegium for Advanced study. He has held research positions at CERN, Harvard, NORDITA and DESY, and served as chair of the High Energy and Particle Physics Division Board of the European Physical Society.

ASYMPTOTIC DIFFRACTION THEORY AND NUCLEAR SCATTERING

ROY J. GLAUBER
Harvard University

PER OSLAND
University of Bergen

CAMBRIDGE
UNIVERSITY PRESS

CAMBRIDGE
UNIVERSITY PRESS

University Printing House, Cambridge CB2 8BS, United Kingdom

One Liberty Plaza, 20th Floor, New York, NY 10006, USA

477 Williamstown Road, Port Melbourne, VIC 3207, Australia

314–321, 3rd Floor, Plot 3, Splendor Forum, Jasola District Centre, New Delhi – 110025, India

79 Anson Road, #06–04/06, Singapore 079906

Cambridge University Press is part of the University of Cambridge.

It furthers the University's mission by disseminating knowledge in the pursuit of education, learning, and research at the highest international levels of excellence.

www.cambridge.org
Information on this title: www.cambridge.org/9781107104112
DOI: 10.1017/9781316219256

First published 2019

Printed in the United Kingdom by TJ International Ltd, Padstow Cornwall

A catalogue record for this publication is available from the British Library.

ISBN 978-1-107-10411-2 Hardback

Contents

Preface

In the first part of this book we review diffraction theory, and introduce its asymptotic form. Whereas diffraction theory rests on two assumptions – (i) a small angle of deflection and (ii) the de Broglie wavelength of the incident particle must be small compared to the size of the target from which it is scattered – the asymptotic form requires a further condition: the wavelength associated with the momentum transfer must also be small compared to the size of the scattering target.

We introduce the concept of "trajectories," which are essential for a semiclassical description. In fact, scattering along a semiclassical trajectory that passes the target nucleus at a certain point in the impact plane will often account for a dominant part of the scattering amplitude. This phenomenon is illustrated by a couple of simple examples. Next, we show that it is useful to extend the trajectory concept to the complex impact parameter plane. This is not only a necessity for scattering situations involving absorption, it is also equally useful for purely nonabsorptive cases, when extended into the angular domain of nonclassical scattering.

In the second part, realizing that the nuclear surface region plays a very important role for the scattered intensity distribution, we systematically compare phase shift functions with Gaussian, exponential and more complicated surface falloffs. For each of the cases of Gaussian and exponential edges, we compare several different phase shift functions having exactly the same edge, but a different internal structure. The importance of singularities of the phase shift functions will be stressed and illustrated. Since the derivative of the phase shift function is central to these considerations, we also present examples where this derivative is the starting point, and the phase shift function itself is constructed by the corresponding integration.

In the third part we discuss composite phase shift functions. The basic example is Coulomb plus nuclear scattering; we supplement this with a discussion of scattering from an extended Coulomb charge. In this case, we show how singularities analogous to rainbows appear at the maximum classically allowed scattering angle. Inside the rainbow, there are often oscillations superimposed on the Rutherford cross section, due to the interference of contributions from two stationary points, whereas outside, in the classically forbidden regime, the cross section falls off in a more exponential fashion.

Furthermore, we discuss sums and differences of two phase shift functions that have different edge falloffs, as well as combinations of two phase shift functions that have different analytic properties, or singularities of different degrees. All these features lead to more complicated intensity distributions than the more regular patterns seen in Part II.

Acknowledgments

The approach to scattering theory presented here was initially explored in collaboration with Marek Bleszynski. We are grateful to him for many suggestions and lively discussions. We are also indebted to Göran Fäldt, with whom we have enjoyed collaborations on closely related topics.

We have furthermore enjoyed, during this work, the hospitality of several remarkable institutions, most noteworthy the Los Alamos National Laboratory, CERN, NORDITA, and ICTP.

Abbreviations

$\boldsymbol{b} = (b_x, b_y)$	impact parameter
$b_{\alpha x}, b_{\beta x}, \ldots$	stationary phase impact parameters
$b_{x,\mathrm{R}}$	impact parameter at rainbow point
c, β	size and diffuseness parameters
$\mathrm{d}\sigma/\mathrm{d}\Omega$	differential cross section
$(\mathrm{d}\sigma/\mathrm{d}\Omega)_{\mathcal{R}}$	differential Rutherford cross section
$f(\boldsymbol{k}', \boldsymbol{k})$	scattering amplitude
$F(b_x)$	scattering amplitude contribution from b_x
F_C	Coulomb scattering amplitude
$\hbar\boldsymbol{k}$	projectile momentum
$\hbar\boldsymbol{q}$	momentum transfer
$\hbar q_{\mathrm{R}}$	momentum transfer at rainbow point
$X(b_x)$	phase shift function for $\boldsymbol{b} = (b_x, 0)$
η	Sommerfeld parameter, see Eq. (4.4)
$\chi(\boldsymbol{b})$	phase shift function

Part I

Introduction

1

Overview and Preview

The beginnings of nuclear physics can be identified quite precisely with the experiments by Rutherford and his coworkers on the elastic scattering of α-particles [19]. While the techniques of measurement and indeed the particles used as projectiles have undergone change many times in the ensuing decades, the fundamental purpose of these elastic scattering measurements remains the same. They have provided much of our understanding of the size and structure of nuclei and of the interactions of elementary particles with them. As the energies of the particles have been raised, these experiments have become more elaborate in technique, but the importance attached to them has grown as well. The fact that the wavelength associated with the incident particles decreases with increasing energy means that measurements made at higher energies are implicitly capable of conveying more detailed information than those made at lower energies. Their resolving power, in other words, increases steadily with energy.

Making measurements at high collision energies characteristically requires the use of instrumentation that is large in scale and ponderous of mass. We need only recall the massive electromagnets required for calibration and momentum analysis of charged particle beams. The reduction and interpretation of the data gathered in these experiments have usually involved analytical problems of corresponding magnitude and weight. The more detailed the experimental measurements have become, the more computation seems to be required in order to extract fundamental data from them. While the most straightforward approaches have often been productive, they have tended to become lengthier with increasing particle energies, and to offer less direct insight into the meaning of even the more prominent features of the accumulated data.

When the kinetic energies of the projectile particles begin to exceed the energies of their interactions with target nuclei, the particles usually suffer only small deflections during collision processes. Since the particle wavelengths, at such energies, are considerably smaller than nuclear dimensions, the angular distributions

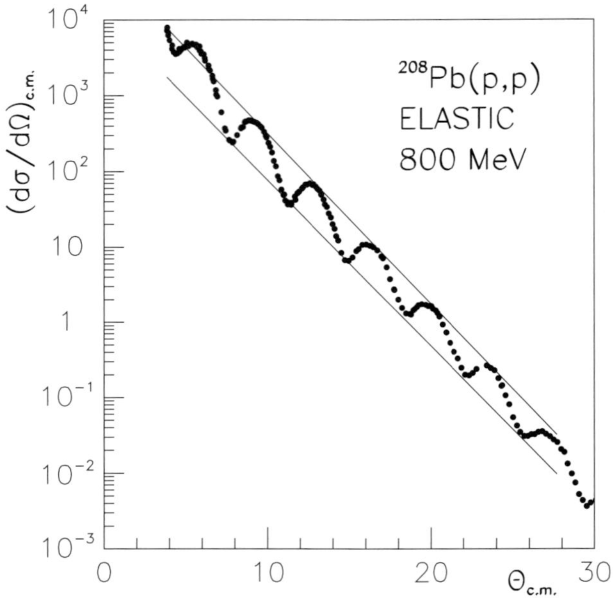

Figure 1.1 Differential cross section for elastic proton-lead scattering at 800 MeV [13].

of elastically scattered particles share many of the familiar properties of optical diffraction patterns. They tend to be strongly peaked for scattering in the forward direction, and to concentrate the scattered intensity within a fairly narrow cone. While the intensity tends to fall rapidly, on the average, with increasing scattering angle, it is also found often to oscillate more or less periodically. These features are quite evident, for example, in Fig. 1.1, which shows a succession of measured values of the differential cross section for the elastic scattering of 800 MeV protons by ^{208}Pb [13]. The measurements, which intentionally omit the high intensity of Coulomb scattering at small angles, show a decrease of the differential cross section by about seven orders of magnitude over the angular range from 3° to 30°. At the same time they show a fairly regular oscillation of the falling intensity which is surely a wave-mechanical interference effect of some sort.

To make accurate measurements of scattering over such a large range of interactions has meant overcoming many experimental challenges [13]. But once they are overcome, the theoretical analysis of the results presents formidable problems as well, and these call for innovative solutions. Our intention is to discuss the means of analyzing such data in the pages that follow by introducing an approximate approach to diffractive scattering, one which deals compactly with broad ranges of intensity. This approach should apply as well to other experiments in which diffracted intensities are studied over a broad range of magnitudes.

We have drawn two straight lines on the differential cross section graph of Fig. 1.1 to show that the average logarithmic slope of the curve is remarkably constant. The successive maxima and minima of the cross section both tend to follow a simple exponential law of decrease. The period of oscillations and their amplitudes both tend to remain surprisingly constant. Such observations immediately lead one to question the value of extending such measurements to still larger scattering angles. Will the angular dependencies to be found there reward the formidable labors of making measurements at extremely low intensities, or will they simply repeat the features of the differential cross section already seen at smaller angles? It would obviously be of value to develop a theory of the scattering process that allows direct insight into the reasons for the behavior noted and says clearly what can be determined by extending the range of measurements.

We shall show that there are indeed certain sorts of changes to be anticipated in the angular distributions at larger angles, and that it may indeed be worth the extensive effort of measuring them. We propose to do this by developing methods of approximation that are related quite closely to those of optical diffraction theory. The time-honored forms of optical diffraction theory are approximations based on integrals that can be evaluated in ways that are straightforward if fairly lengthy, but their results are not always transparent. We do not propose here to improve the accuracy of this approach. What we shall do is to discuss further approximations that give qualitatively clear insights into the nature of the results, and ones that often present no great sacrifice of accuracy. These methods furnish a considerable amount of analytic as well as numerical insight, we find, and engender qualitative understandings that have not been provided by the lengthier and less accessible, if more exact, methods of solving the scattering problem.

There are at least two traditional methods of calculating angular distributions of nuclear scattering. The most familiar depends on the separation of angular and radial variables. If the scattering interaction is spherical, the orbital angular momentum of the incident particle is conserved, and the Schrödinger equation can be separated into a sequence of radial differential equations, one for each integer value of the angular momentum. In that case integrating each of the radial equations leads to the evaluation of a complex phase shift for that angular momentum, and these in turn contribute to an explicit expression for the scattering amplitude as a sum of spherical harmonic functions (see Appendix A).

Though this procedure for evaluating the scattering amplitude is implicitly exact, it encounters serious difficulties in treating the problems we are considering. Measurements of scattering are most informative when the wavelength λ of the incident particle is much smaller than the radius R of the scatterer. In that case the number of radial equations to be integrated must be at least as large as R/λ, the minimal number of partial waves affected by this scattering. In practice it must often be much larger, particularly if the nuclear interaction does not cut off sharply at radius R. So there is no alternative to calculating a great many phase shifts. But there is a further difficulty implicit in the summation of the partial wave contributions. As we can see in the seven orders of magnitude by which the cross section drops in Fig. 1.1, there must be wholesale cancellation of the amplitudes contributed by the different partial waves. In the actual experiments, nature provides that cancellation effortlessly, but for calculational purposes, it places extreme demands on the accuracy with which the individual phase shifts must be determined. If compactly interpretable results are desired, this exact route would not be the one to follow.

An alternative approach, one that deals well with sums over large numbers of partial waves, corresponds closely to optical diffraction theory. If we are only interested in scattering through relatively small angles, we can express the scattering amplitude as an integral over the plane of impact parameters very much as it is done in Fraunhofer diffraction theory [21]. Evaluation of the scattering amplitudes is thus reduced to numerical evaluation of certain Fourier integrals. But here too the numerical approach encounters prohibitive problems. It would not be easy to maintain the accuracy required to reach a meaningful result given all the cancellation that must be implicit in the drop of the cross section by many orders of magnitude.

We conclude that it is a better strategy to isolate, if we can, the asymptotic behaviors of these scattering amplitudes, and to study them more directly, in order to interpret what goes on at larger momentum transfers. This approach, as we shall see, will involve further approximations, but with small and controllable errors. It will lead us gradually away from the analogies with classical diffraction theory with which our analysis begins, and introduce us to the discussion of particle trajectories in a space in which the particle coordinates become complex, rather than the familiar one in which they remain real-valued. It will reveal particle behaviors that depart significantly from those of classical mechanics and describe new behaviors of scattering amplitudes. Our hope is thereby to be able to find expressions for the amplitudes that are both compact and suggestive of new interpretations.

As a way of illustrating some of the properties of diffraction theory, let us consider the transmission of a plane light wave of propagation vector $\boldsymbol{k}$, $(k = 2\pi/\lambda)$, through an aperture in a flat opaque screen. We take the screen to define the plane of coordinate vectors $\boldsymbol{b}$ and assume that the amplitude of the scalar wave at any point in the plane of the aperture is $A(\boldsymbol{b})$. Then, within the approximations of Fraunhofer diffraction theory, the amplitude $f(\boldsymbol{k}', \boldsymbol{k})$ of the wave diffracted in the direction $\boldsymbol{k}'$ $(k' = k)$ is proportional to the two-dimensional Fourier integral [8],

$$f(\boldsymbol{k}', \boldsymbol{k}) \sim \int e^{i(\boldsymbol{k}-\boldsymbol{k}')\cdot\boldsymbol{b}}\, A(\boldsymbol{b})\, \mathrm{d}^2 b, \tag{1.1}$$

taken over the plane of the aperture, with $\mathrm{d}^2 b$ the element of area.

The slits and pinholes of physical optics have usually been made of metal. Their hard-edged character has meant that $|A(\boldsymbol{b})|$ is approximated as decreasing abruptly to zero at the edge of the aperture. We could, in principle however, fabricate (for example, by photographic means) an aperture such as a slit of continuously varying transparency, which transmits a wave with a smooth intensity profile. Let us consider, for example, a straight slit parallel to the y-axis, which transmits the smoothly varying wave amplitude

$$A(x) = \frac{\beta}{\pi} \frac{1}{x^2 + \beta^2}, \tag{1.2}$$

where x is the coordinate transverse to the slit and β is a measure of its width. Then the amplitude of the diffracted wave can be written as the integral

$$f(\boldsymbol{k}', \boldsymbol{k}) \sim \frac{\beta}{\pi} \int_{-\infty}^{\infty} \frac{e^{-iqx}}{x^2 + \beta^2} \, \mathrm{d}x \tag{1.3}$$

$$\sim \frac{1}{2\pi i} \int_{-\infty}^{\infty} e^{-iqx} \left\{ \frac{1}{x - i\beta} - \frac{1}{x + i\beta} \right\} \mathrm{d}x, \tag{1.4}$$

where $\boldsymbol{q} = \boldsymbol{k}' - \boldsymbol{k}$. The integrand obviously has two poles in the complex x-plane. For $q < 0$ the contour can be closed by a semicircle in the upper half-plane, and then the pole at $x = i\beta$ contributes the residue $e^{q\beta}$. For $q > 0$ the contour can be closed in the lower half-plane and the residue is $e^{-q\beta}$, but the contour is closed in a clockwise sense. The result for both cases may be stated in the form

$$f(\boldsymbol{k}', \boldsymbol{k}) \sim e^{-\beta|\boldsymbol{q}|}. \tag{1.5}$$

Since $|\boldsymbol{q}| = |\boldsymbol{k}' - \boldsymbol{k}| = 2k \sin \frac{\theta}{2}$, and at small angles this is just $k\theta$, we already have a hint of how exponential decreases of intensity, such as that shown in Fig. 1.1, can arise.

There are in fact two lessons to be learned here. The first is that a smooth absorption profile leads to a much more rapid decrease of intensity with increasing scattering angle than is familiar in optical experiments with sharp-edged apertures. The second is somewhat more abstract. For $q > 0$, for example, a single, complex value of x, the pole at $x = -i\beta$, furnishes a scattering amplitude equivalent to the whole range of real values of x that span the width of the slit. This is an elementary illustration of a technique of using complex trajectories that we shall explore further.

The scattered intensity $|f(\boldsymbol{k}', \boldsymbol{k})|^2$ for our diffuse slit drops off exponentially, according to Eq. (1.5), but it does not show any of the oscillations evident in Fig. 1.1. To secure such oscillations we need only let the waves transmitted by two such slits interfere [3,15]. Let us assume, for example, that we have two such slits that are parallel to each other and centered at $x = \pm c$. Then the wave amplitudes transmitted by the screen can be written according to the integral Eq. (1.1), which represents the approximation of Fraunhofer diffraction theory, as

$$A(x) = \frac{\beta}{2\pi}\left\{\frac{1}{(x-c)^2+\beta^2} + \frac{1}{(x+c)^2+\beta^2}\right\}, \tag{1.6}$$

and the corresponding diffraction amplitude, according to Eqs. (1.5) and (1.6), must then be

$$f(\boldsymbol{k}', \boldsymbol{k}) \sim e^{-\beta|\boldsymbol{q}|} \cos(cq). \tag{1.7}$$

The angular separation of the successive minima in the intensity is then fixed and varies inversely with the distance between the slits. There are several ways in which the diffraction pattern corresponding to Eq. (1.7) still differs from the angular distribution of Fig. 1.1. It has periodic zeros while the experimental intensity oscillates between positive bounds. Furthermore, the phase of its oscillations is somewhat different. Both of these differences can be addressed by assuming that some weak refraction takes place as the wave penetrates the aperture, and therefore adding appropriate phase shifts to the amplitude $A(\boldsymbol{b})$.

This set of ad hoc assumptions can readily be expanded to account for more of the features of the angular distributions measured in elastic nuclear scattering, but to proceed further by such means would risk omitting a number of interesting physical phenomena. It would furthermore leave the geometrical basis of the method quite unclear. Instead we shall begin in the next chapter the development of a systematic analysis that yields a more comprehensive version of the double-slit paradigm for proton scattering by lead and for a broad range of other cases that have been measured experimentally. But the two-slit paradigm is not universal in its application and, as we shall see, there are many cases in which the systematic analysis yields amplitudes that correspond to single-slit patterns or to those of multiplicities of dissimilar slits. The key to our analysis will be the application of a simple asymptotic approximation to the theory of Fraunhofer diffraction. We shall show, through this approach, that it is possible to give simple explanations for a considerable variety of behaviors of the differential cross section and shall discuss explicitly many of the more interesting behaviors that have been, or are likely to be, observed. We hope that the various cases we have treated can thus provide something akin to an atlas of the simpler diffractive angular distributions.

Regularities similar to those we have noted in elastic scattering are also present in the angular distributions of inelastic scattering. Three examples that occur in the scattering of protons by ^{144}Sm at 800 MeV [7] are shown in Fig. 1.2. These distributions too show a nearly constant average logarithmic slope and more or less regularly spaced oscillations. These features are present in inelastic scattering, in fact, for nearly the same reasons that cause their presence in elastic scattering.

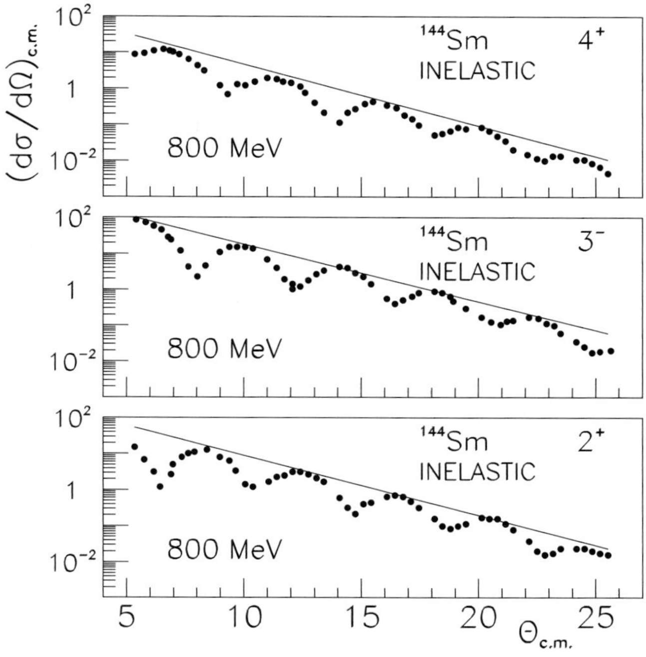

Figure 1.2 Differential cross section for inelastic proton-samarium scattering at 800 MeV [7].

Whatever rearrangement of nucleons an incident particle may produce in an inelastic collision, the collision process may be pictured as taking place against a background of elastic scattering by the nucleus as a whole. It is important therefore to concentrate on the description of elastic scattering, before moving on to the consideration of inelastic scattering.

2

Diffraction Theory

There are two principal assumptions that underlie Fraunhofer diffraction theory. The first of these is that the wavelength of the incident field be much smaller than the size of any aperture through which it passes. In the three-dimensional scattering problems that interest us, this becomes the condition that the wavelength be smaller than the range of any of the forces that deflect the incident particle. A second condition is less familiar in the optical context, where refractive effects are often negligible or even absent, but it is equally important for more general sorts of scattering processes. It is that the kinetic energy of the incident particle be sufficiently high that the force fields that it encounters tend to lead mainly to small deflections.

Under these conditions nearly all of the scattered intensity is projected into a diffraction pattern close to the forward direction. If $\boldsymbol{k}$ and $\boldsymbol{k}'$ are the initial and final propagation vectors of the projectile wave function, or alternatively the initial and final momenta divided by $\hbar$, the momentum transfers $\hbar\boldsymbol{q} = \hbar(\boldsymbol{k}' - \boldsymbol{k})$ will tend characteristically to be small in magnitude, $q/k \ll 1$. The distribution of these momentum transfers may nonetheless be quite informative about the force fields responsible for the scattering processes.

As long as our two assumptions hold true, incident particles will tend to pass through the region of interaction with little deflection, that is to say they will continue, at least approximately, to follow straight-line paths through the interaction region. Let us assume that the incident projectile is described by the plane wave $e^{i\boldsymbol{k}\cdot\boldsymbol{r}}$ and that the nuclear force can be described for the present by a static potential of some sort. Then in the region lying in the "shadow" of the potential, i.e., for points lying beyond the potential, but not too far from it, the wave function can be approximated as $\exp[i\boldsymbol{k}\cdot\boldsymbol{r} + i\chi(\boldsymbol{b})]$, where $\boldsymbol{b}$ is an impact vector, the component of $\boldsymbol{r}$ perpendicular to $\boldsymbol{k}$, and χ is a phase shift function that describes the cumulative effect of passage through the potential.

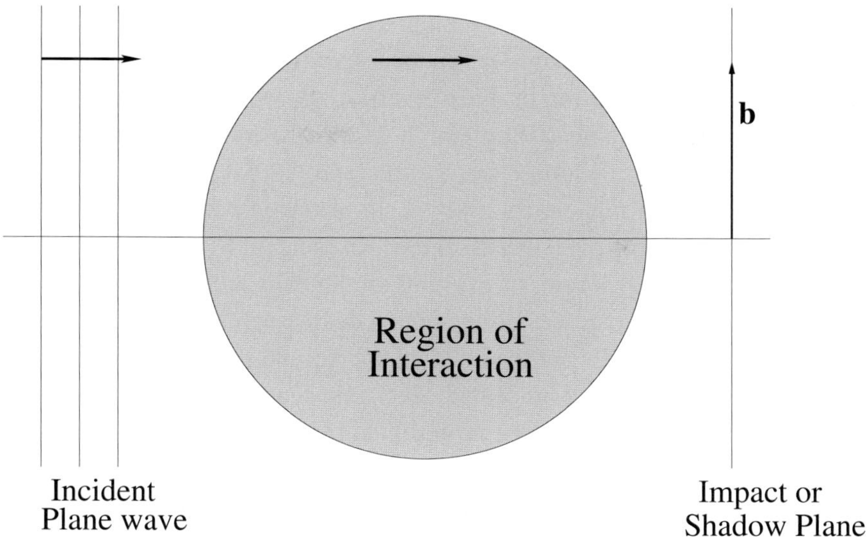

Figure 2.1 Geometry of the scattering process.

It is convenient to imagine constructing a shadow or impact plane perpendicular to $\boldsymbol{k}$ in the region immediately shadowed by the nuclear interaction, i.e., close to the nucleus but just beyond its range of interaction, so that the lateral spreading of the wave function components may remain quite small. (See Fig. 2.1.) If we take the wave function in that plane to be $\exp[i\boldsymbol{k}\cdot\boldsymbol{r}+i\chi(\boldsymbol{b})]$ and then apply Green's theorem to evaluate the wave function in the half-space beyond it, we can find the scattering amplitude from the form it takes far from the plane. For small momentum transfers, that scattering amplitude is given by [8]

$$f(\boldsymbol{k}',\boldsymbol{k}) = \frac{ik}{2\pi}\int e^{-i\boldsymbol{q}\cdot\boldsymbol{b}}\left\{1 - e^{i\chi(\boldsymbol{b})}\right\} \mathrm{d}^2 b. \tag{2.1}$$

The integration over impact vectors $\boldsymbol{b}$ in this expression can be thought of as being carried out over the impact plane. The scattering amplitude Eq. (2.1) is in fact a slightly generalized form of the familiar Kirchhoff integral for the amplitude of Fraunhofer diffraction. It possesses a number of derivations other than the one we have outlined [8], and each of these seems to suggest other generalizations, or to have mathematical advantages of its own. We shall confine ourselves here simply to mentioning several of them.

When the function $\chi(\boldsymbol{b})$ has azimuthal symmetry, for example, about the direction $\boldsymbol{k}$, i.e., when $\chi = \chi(b)$ the angular integral in Eq. (2.1) may be expressed in terms of the Bessel function J_0. We then have [1]

$$f(\boldsymbol{k}',\boldsymbol{k}) = ik\int_0^\infty J_0(qb)\left\{1 - e^{i\chi(b)}\right\} b\,\mathrm{d}b, \tag{2.2}$$

an integral that may be regarded as the high-energy limit of the general expression for a scattering amplitude as a sum of spherical harmonic functions [12,16,20]. We have included a brief discussion of this expansion and its relation to the diffraction theory in Appendix A. We need only note here that the impact parameter b corresponds to the angular momentum ℓ (for ℓ sufficiently large) through $kb = \ell + \frac{1}{2}$.

The representation of nuclear forces by a simple static potential would, if taken literally, be a considerable oversimplification. An incident high energy particle has an appreciable chance, as a rule, of exciting the nucleus and leaving some part of its energy behind. Such processes bring about a decrease in amplitude of the part of the wave function propagating at the incident energy and it is from that part that we calculate the elastic scattering amplitude. Inelastic scattering, in other words, is a loss mechanism and its effect can be represented by letting the phase shift function $\chi(b)$ have a positive imaginary part. (That representation amounts, in the simplest optical diffraction experiments, to letting $\exp\{i\chi(\boldsymbol{b})\}$ be zero wherever an absorbing screen obstructs the wave, and letting it equal unity within any open aperture.)

It would go far beyond our present needs to delve into the theory of nuclear forces, but a further element of generality present in Eq. (2.1) is surely worth mentioning. The representation of the nuclear interaction by means of a complex phase shift function $\chi(\boldsymbol{b})$ is considerably more general than its description by means of a potential $V(\boldsymbol{r})$, be it real or complex. Not all field interactions can be described accurately by means of a local potential, but under the assumptions we have stated there must be a phase shift function $\chi(\boldsymbol{b})$, whether there exists such a local potential or not.

If the interaction is indeed describable by means of a local scalar potential $V(\boldsymbol{r})$ then, within the context of the Klein–Gordon equation, the phase shift function is given approximately by [8]

$$\chi(\boldsymbol{b}) = -\frac{1}{\hbar v}\int_{-\infty}^{\infty} V(\boldsymbol{b} + \hat{\boldsymbol{\kappa}} z)\, \mathrm{d}z \tag{2.3}$$

where $\hat{\boldsymbol{\kappa}}$ is the unit vector $\hat{\boldsymbol{\kappa}} = \boldsymbol{k}/k$ and v is the velocity of the incident particle. If V is a spherically symmetric potential, $V = V(r)$, this expression reduces to

$$\chi(b) = -\frac{1}{\hbar v}\int_{-\infty}^{\infty} V\left(\sqrt{b^2 + z^2}\right) \mathrm{d}z \tag{2.4}$$

$$= -\frac{2}{\hbar v}\int_{b}^{\infty} \frac{V(r) r\, \mathrm{d}r}{\sqrt{r^2 - b^2}} \tag{2.5}$$

If the analysis of experimental data were somehow to furnish us with an azimuthally symmetric phase shift function $\chi(b)$ we could then regard Eq. (2.5) as an integral equation to determine a corresponding local and spherically symmetric potential $V(r)$. The integral equation, Eq. (2.5), is in fact of the Abel variety and possesses the explicit solution [8]

$$V(r) = \frac{\hbar v}{\pi r}\frac{\mathrm{d}}{\mathrm{d}r}\int_r^\infty \frac{\chi(b) b\,\mathrm{d}b}{\sqrt{b^2 - r^2}} \tag{2.6a}$$

$$= \frac{\hbar v}{\pi}\frac{\mathrm{d}}{\mathrm{d}(r^2)}\int_{-\infty}^\infty \chi\left(\sqrt{r^2+z^2}\right)\mathrm{d}z \tag{2.6b}$$

$$= \frac{\hbar v}{2\pi}\int_{-\infty}^\infty \frac{\chi'\left(\sqrt{r^2+z^2}\right)}{\sqrt{r^2+z^2}}\,\mathrm{d}z. \tag{2.6c}$$

This is, in effect, an "optical potential," equivalent to the phase shift function $\chi(b)$ in the sense that it leads to the same scattering amplitude. It is often unclear, however, whether the "optical potential" has any physical meaning deeper than that. In the present approach, therefore, we shall give considerably greater emphasis to the phase shift function $\chi(\boldsymbol{b})$ which stands, after all, in a much closer relation to experimental observation.

A further point concerning the approximate scattering amplitude, Eq. (2.1), is worth noting. A number of analyses have shown that the angular range of the approximation can be extended somewhat [22] by taking the vector $\hat{\boldsymbol{\kappa}}$ to be in the direction of the vector $\frac{1}{2}(\boldsymbol{k}+\boldsymbol{k}')$, rather than $\boldsymbol{k}$, and by taking the impact plane to be perpendicular to it. We have illustrated this choice in Fig. 2.2, which shows $\hat{\boldsymbol{\kappa}} = (\boldsymbol{k}+\boldsymbol{k}')/|\boldsymbol{k}+\boldsymbol{k}'|$ and the two auxiliary vectors, $\hat{\boldsymbol{q}} = (\boldsymbol{k}'-\boldsymbol{k})/|\boldsymbol{k}'-\boldsymbol{k}|$ and

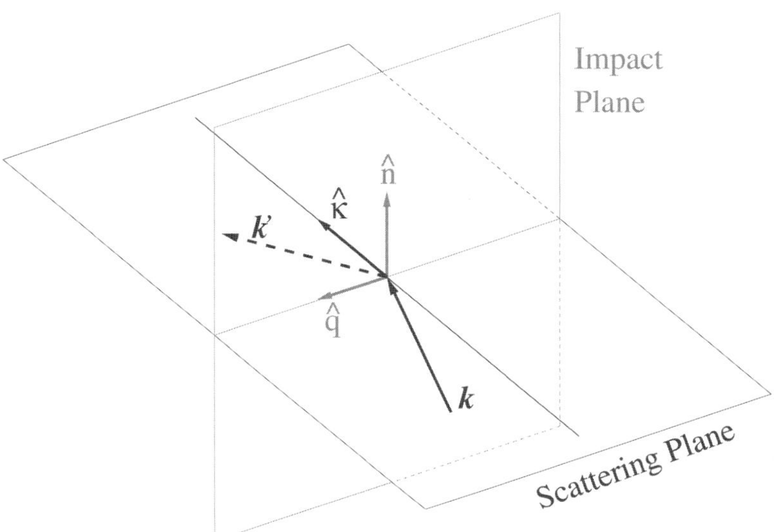

Figure 2.2 Coordinate system used.

$\hat{\boldsymbol{n}} = \boldsymbol{k} \times \boldsymbol{k}'/|\boldsymbol{k} \times \boldsymbol{k}'|$. These are mutually orthogonal unit vectors. They have been defined to form a right-handed triad in the sense that $\hat{\boldsymbol{q}} \times \hat{\boldsymbol{n}} = \hat{\boldsymbol{\kappa}}$.[1]

Scattering amplitudes such as that of Eq. (2.1) possess several symmetry properties that will be useful to us at later points. They are invariant under velocity reversal,

$$f(-\boldsymbol{k}, -\boldsymbol{k}') = f(\boldsymbol{k}', \boldsymbol{k}), \tag{2.7}$$

a relation that is equivalent to "time reversal" symmetry, an invariance under reversal of both the order and the signs of the initial and final momenta. It holds only when the phase shift function $\chi(\boldsymbol{b})$ is real-valued. Furthermore, for interactions that are also invariant under coordinate inversion we have $\chi(-\boldsymbol{b}) = \chi(\boldsymbol{b})$ and consequently

$$f(-\boldsymbol{k}, -\boldsymbol{k}') = f(\boldsymbol{k}, \boldsymbol{k}') \tag{2.8}$$

so that the scattering amplitude according to Eq. (2.7) then has the symmetry

$$f(\boldsymbol{k}', \boldsymbol{k}) = f(\boldsymbol{k}, \boldsymbol{k}'). \tag{2.9}$$

The scattering amplitude, Eq. (2.1), also possesses an interesting symmetry that is quite specific to the diffraction approximation from which it stems. If we write the amplitude $f(\boldsymbol{k}', \boldsymbol{k})$ as a function of the phase shift function χ and the vector $\boldsymbol{q}$, i.e.,

$$f(\boldsymbol{k}', \boldsymbol{k}) = \mathcal{F}[\chi, \boldsymbol{q}],$$

then the transformation $\boldsymbol{q} \to -\boldsymbol{q}$ and $\chi \to -\chi^*$ (which preserves the absorptive character of the phase shift function) carries the scattering amplitude into its complex conjugate and changes its sign,

$$\mathcal{F}[-\chi^*, -\boldsymbol{q}] = -\mathcal{F}^*[\chi, \boldsymbol{q}]. \tag{2.10}$$

In particular then, if the phase shift function is symmetric under spatial inversion, and furthermore purely imaginary, the scattering amplitude obeys the relations

$$\mathcal{F}[\chi, -\boldsymbol{q}] = \mathcal{F}[\chi, \boldsymbol{q}] = -\mathcal{F}^*[\chi, \boldsymbol{q}], \tag{2.11}$$

which show it must then be purely imaginary.

[1] The sign of $\boldsymbol{q}$ is opposite to that used in [3].

3

Asymptotic Diffraction Theory and Classical Scattering

It is quite instructive, as a first step in exploring the asymptotic behavior of the differential scattering amplitude, to observe the way in which it approaches its classical limit for real-valued interactions. We can simplify the expression in Eq. (2.1) a bit for that purpose by noting that it contains the two-dimensional delta function[1]

$$\frac{1}{2\pi}\int e^{-i\boldsymbol{q}\cdot\boldsymbol{b}}\,\mathrm{d}^2 b = 2\pi\,\delta^{(2)}(\boldsymbol{q}), \tag{3.1}$$

which is known when defined with suitable care to vanish for $\boldsymbol{q} \neq 0$. To evaluate the integral Eq. (2.1) for the scattering amplitude, in other words, we will need only to fix our attention on the remaining term, the two-dimensional Fourier transform of $\exp\{i\chi(\boldsymbol{b})\}$, and to take the same care in carrying out the required integration. Neither the expression Eq. (3.1) for the delta function, nor the Fourier transform of $\exp\{i\chi(\boldsymbol{b})\}$ is a properly convergent integral, however, until we proceed a bit more cautiously by defining them, for example, by means of an appropriate summation convention.

For interactions of finite range, for example, we do not change the scattering amplitude $f(\boldsymbol{k}',\boldsymbol{k})$ of Eq. (2.1) by defining it as the limit

$$f(\boldsymbol{k}',\boldsymbol{k}) = \lim_{\epsilon\to 0}\frac{ik}{2\pi}\int e^{-\epsilon b^2 - i\boldsymbol{q}\cdot\boldsymbol{b}}\left\{1 - e^{i\chi(\boldsymbol{b})}\right\}\mathrm{d}^2\boldsymbol{b}, \tag{3.2}$$

taken as ϵ goes to zero through positive values. While this limiting procedure does not change the scattering amplitude, it does give a unique definition for arbitrarily

[1] The Dirac delta function [6] $\int \mathrm{d}x\, \exp(-iqx) = 2\pi\delta(q)$ is defined as vanishing save in the neighborhood of $q = 0$, and yet providing the integral $\int \mathrm{d}q\,\delta(q) = 1$. We have used the symbol $\delta^{(2)}(\boldsymbol{q})$ for the two-dimensional version of the function defined on the $\boldsymbol{q}$-plane.

small ϵ to the two terms that make it up. In particular, the delta function in Eq. (3.1) is replaced by

$$\frac{1}{2\pi}\int e^{-\epsilon b^2 - i\boldsymbol{q}\cdot\boldsymbol{b}}\,\mathrm{d}^2\boldsymbol{b} = \frac{1}{2\epsilon}e^{-\frac{q^2}{4\epsilon}}, \tag{3.3}$$

which becomes vanishingly small for $q \gg 2\sqrt{\epsilon}$. The scattering amplitude for $q \gg 2\sqrt{\epsilon}$ is then given by the well-defined integral

$$f(\boldsymbol{k}',\boldsymbol{k}) = \lim_{\epsilon\to 0}\frac{k}{2\pi i}\int \exp\{-\epsilon b^2 - i\boldsymbol{q}\cdot\boldsymbol{b} + i\chi(\boldsymbol{b})\}\,\mathrm{d}^2\boldsymbol{b}. \tag{3.4}$$

In the classical limit, $\hbar \to 0$, the de Broglie wavelength of the incident particle goes to zero. The magnitude of the vector $\boldsymbol{q} = \boldsymbol{k}' - \boldsymbol{k}$ for any fixed angle of scattering then becomes infinite. The phase shift $\chi(b)$, as we can see from Eq. (2.3), also becomes infinite in magnitude, and may be expected to vary rapidly with the position $\boldsymbol{b}$ as well. The principal contributions to the integral in Eq. (3.4) are bound therefore to come from the immediate neighborhoods of the points at which the phase $-\boldsymbol{q}\cdot\boldsymbol{b} + \chi(\boldsymbol{b})$ is stationary. These are the points in the impact plane which satisfy the relation

$$\nabla_b\{-\boldsymbol{q}\cdot\boldsymbol{b} + \chi(\boldsymbol{b})\} = 0$$

or

$$\boldsymbol{q} = \nabla_b\chi(\boldsymbol{b}). \tag{3.5}$$

The classical significance of this stationary phase condition is immediately clear. If a particle at the position $\boldsymbol{r} = \boldsymbol{b} + \hat{\boldsymbol{\kappa}}z$ is subject to a potential $V(\boldsymbol{b} + \hat{\boldsymbol{\kappa}}z)$ then it experiences a transverse force $-\nabla_b V(\boldsymbol{b} + \hat{\boldsymbol{\kappa}}z)$. The integral of this force over time, according to Eq. (2.3), is given by $\hbar\nabla_b\chi(\boldsymbol{b})$. That transverse impulse then must represent the transfer of momentum to the scattered particle,

$$\hbar(\boldsymbol{k}' - \boldsymbol{k}) = \hbar\boldsymbol{q} = \hbar\nabla_b\chi(\boldsymbol{b}). \tag{3.6}$$

This is precisely the stationary phase condition in Eq. (3.5). It possesses only a discrete set of roots for the impact vector $\boldsymbol{b}$. There is, in other words, at most only a discrete set of classical trajectories, if indeed there are any at all, that can lead to scattering for any given momentum transfer $\hbar\boldsymbol{q}$.

To find the roots of Eq. (3.5) it is convenient to adopt a coordinate system within the impact plane. We can use the unit vectors $\hat{\boldsymbol{q}}$ and $\hat{\boldsymbol{n}}$ defined in connection with Fig. 2.2 to write

$$\boldsymbol{b} = \hat{\boldsymbol{q}}b_x + \hat{\boldsymbol{n}}b_y, \tag{3.7}$$

and express the phase shift as a function of its Cartesian coordinates, $\chi(b_x, b_y)$. The stationary points are then determined by the pair of equations

$$\frac{\partial}{\partial b_x}\chi(b_x, b_y) = q \tag{3.8}$$

$$\frac{\partial}{\partial b_y}\chi(b_x, b_y) = 0. \tag{3.9}$$

The phase shift functions we encounter most often are rotationally invariant,

$$\chi(\boldsymbol{b}) = \chi(b) = \chi\left(\sqrt{b_x^2 + b_y^2}\right), \tag{3.10}$$

and in that case Eqs. (3.8) and (3.9) reduce to

$$\frac{b_x}{b}\chi'(b) = q \tag{3.11a}$$

$$\frac{b_y}{b}\chi'(b) = 0. \tag{3.11b}$$

The latter of these equations shows that either b_y or $\chi'(b)$ must be zero. If $\chi'(b)$ were to vanish, however, there could be no stationary point for finite q according to Eq. (3.11a). A stationary point for $q \neq 0$, therefore, can only occur for $b_y = 0$, i.e., for $\boldsymbol{b}$ lying in the scattering plane. The coordinate b_x is then determined by what remains of Eq. (3.11),

$$\frac{b_x}{|b_x|}\chi'(|b_x|) = q. \tag{3.12}$$

Of course, when the phase shift function lacks this rotational symmetry, the stationary points could lie anywhere in the impact plane.

We can approximate the scattering amplitude, Eq. (3.4), by expanding the phase shift function in its integrand in the neighborhood of its stationary points. If $\boldsymbol{b}_0$ is a stationary point, for example, we can write, in dyadic notation (i.e., double scalar product)

$$\chi(\boldsymbol{b}) - \boldsymbol{q}\cdot\boldsymbol{b} = \chi(\boldsymbol{b}_0) - \boldsymbol{q}\cdot\boldsymbol{b}_0 + \tfrac{1}{2}(\boldsymbol{b}-\boldsymbol{b}_0)(\boldsymbol{b}-\boldsymbol{b}_0) : \nabla_b\nabla_b\chi(\boldsymbol{b})\Big|_{\boldsymbol{b}_0}. \tag{3.13}$$

For the special case of rotational symmetry $\chi(\boldsymbol{b}) = \chi(b)$ we have

$$\nabla_b\nabla_b\chi(\boldsymbol{b}) = \frac{\boldsymbol{b}}{b}\frac{\boldsymbol{b}}{b}\chi''(b) - \frac{\boldsymbol{b}\boldsymbol{b}}{b^3}\chi'(b) + \frac{\mathbb{I}}{b}\chi'(b), \tag{3.14}$$

where $\mathbb{I}$ stands for the unit dyadic, $\hat{\boldsymbol{q}}\hat{\boldsymbol{q}} + \hat{\boldsymbol{n}}\hat{\boldsymbol{n}}$. It is convenient to introduce Cartesian coordinates centered at the stationary point $\boldsymbol{b}_0$ by writing

$$\boldsymbol{b} - \boldsymbol{b}_0 = \hat{\boldsymbol{q}}x + \hat{\boldsymbol{n}}y \tag{3.15}$$

so that we have

$$\tfrac{1}{2}(\boldsymbol{b}-\boldsymbol{b}_0)(\boldsymbol{b}-\boldsymbol{b}_0):\nabla_b\nabla_b\chi(\boldsymbol{b})\Big|_{\boldsymbol{b}_0}=\alpha_x x^2+\alpha_y y^2, \tag{3.16}$$

and the phase function can be written as

$$\chi(\boldsymbol{b})-\boldsymbol{q}\cdot\boldsymbol{b}=\chi(\boldsymbol{b}_0)-\boldsymbol{q}\cdot\boldsymbol{b}_0+\alpha_x x^2+\alpha_y y^2, \tag{3.17}$$

where

$$\alpha_x \equiv \tfrac{1}{2}\chi''(|b_{0x}|) \tag{3.18a}$$

$$\alpha_y = \chi'(|b_{0x}|)/2|b_{0x}| = q/2b_{0x}. \tag{3.18b}$$

The contribution of the stationary point $\boldsymbol{b}_0$ to the scattering amplitude, Eq. (3.4), can now be written as

$$f_0(\boldsymbol{k}',\boldsymbol{k})=\frac{k}{2\pi i}\,e^{-i\boldsymbol{q}\cdot\boldsymbol{b}_0+i\chi(b_0)}\,\mathcal{I}_0(\boldsymbol{q}), \tag{3.19}$$

where

$$\mathcal{I}_0(\boldsymbol{q})=\lim_{\epsilon\to 0}\int e^{-\epsilon(x^2+y^2)+i\alpha_x x^2+i\alpha_y y^2}\,\mathrm{d}x\,\mathrm{d}y \tag{3.20}$$

is a Gaussian integral to be carried out over the impact plane. Its two Cartesian factors are given by

$$\lim_{\epsilon\to 0}\int_{-\infty}^{\infty} e^{-(\epsilon-i\alpha_y)y^2}\,\mathrm{d}y=\left(\frac{i\pi}{\alpha_y}\right)^{\frac{1}{2}} \tag{3.21a}$$

and

$$\lim_{\epsilon\to 0}\int_{-\infty}^{\infty} e^{-(\epsilon-i\alpha_x)x^2}\,\mathrm{d}x=\left(\frac{i\pi}{\alpha_x}\right)^{\frac{1}{2}}. \tag{3.21b}$$

The integral $\mathcal{I}_0(\boldsymbol{q})$ is therefore given by the product

$$\begin{aligned}\mathcal{I}_0(\boldsymbol{q}) &= \frac{\pi}{(-\alpha_x\alpha_y)^{\frac{1}{2}}}\\ &= 2\pi\left(\frac{-b_{0x}}{\chi'(b_{0x})\chi''(b_{0x})}\right)^{\frac{1}{2}}.\end{aligned} \tag{3.22}$$

The contribution of the assumed stationary point $\boldsymbol{b}_0$ to the scattering amplitude, Eq. (3.4), can thus be written as

$$f_0(\boldsymbol{k}',\boldsymbol{k})=\frac{k}{i}\left(\frac{-b_{0x}}{q\chi''(b_{0x})}\right)^{\frac{1}{2}}e^{-iqb_{0x}+i\chi(b_{0x})} \tag{3.23}$$

where we have made use of Eq. (3.11) to substitute q/b_{0x} for $\chi'(b_{0x})/b_{0x}$. To avoid ambiguities of sign for the actual physical scattering amplitude we must be careful to specify which branch of the square root factor is to be used in this expression. We take it to be the branch for which $\{e^{i\theta}\}^{\frac{1}{2}} = e^{\frac{1}{2}i\theta}$ for $-\pi < \theta \leq \pi$. The scattering amplitude must retain the diffraction symmetry we have stated in Eq. (2.10) and that furnishes an important check on the consistency of the phases in Eq. (3.23).

If we do not assume rotational symmetry it is necessary to integrate the exponential function of a more general quadratic form, but that can easily be done and yields the more general result

$$f_0(\boldsymbol{k}', \boldsymbol{k}) = \frac{k}{\left\{\det \nabla_b \nabla_b \chi(\boldsymbol{b})\big|_{\boldsymbol{b}_0}\right\}^{\frac{1}{2}}} \exp\{i[-\boldsymbol{q} \cdot \boldsymbol{b}_0 + \chi(\boldsymbol{b}_0)]\}. \tag{3.24}$$

The determinant in this expression is that of the matrix of second derivatives of the phase shift function $\chi(\boldsymbol{b})$ evaluated at the stationary point $\boldsymbol{b}_0$.

It is quite interesting to compare this quantum-mechanical result with the classical cross section. To find the classical cross section we need only know the element of solid angle into which the incident particle is projected when it impinges on the element of area $\mathrm{d}\sigma = \mathrm{d}^2\boldsymbol{b}$ of the impact plane in the neighborhood of $\boldsymbol{b}$. The momentum transfer $\hbar q$ and the impact vector $\boldsymbol{b}$ are related by the classical Eq. (3.6) (which is also the stationary phase condition). It follows then that an element of area $\mathrm{d}^2\boldsymbol{b}$ is related to a two-dimensional element of $\boldsymbol{q}$-vectors via

$$\mathrm{d}^2\boldsymbol{b} = \frac{\partial(\boldsymbol{b})}{\partial(\boldsymbol{q})}\, \mathrm{d}^2\boldsymbol{q} \tag{3.25}$$

in which $\partial(\boldsymbol{b})/\partial(\boldsymbol{q})$ is a Jacobian determinant,

$$\frac{\partial(\boldsymbol{b})}{\partial(\boldsymbol{q})} = \frac{1}{\left|\det \nabla_b \nabla_b \chi(\boldsymbol{b})\right|_{\boldsymbol{b}_0}} \tag{3.26}$$

evaluated at the point $\boldsymbol{b}_0$. Since the element of solid angle $\mathrm{d}\Omega$ is given near the forward direction by $\mathrm{d}^2\boldsymbol{q} = k^2\,\mathrm{d}\Omega$, we find that the classical differential cross section is

$$\frac{\mathrm{d}\sigma_0}{\mathrm{d}\Omega} = k^2 \frac{\partial(\boldsymbol{b})}{\partial(\boldsymbol{q})}\bigg|_{\boldsymbol{b}_0} \tag{3.27a}$$

$$= \frac{k^2}{|\det \nabla_b \nabla_b \chi(\boldsymbol{b})|_{\boldsymbol{b}_0}}. \tag{3.27b}$$

This is, of course, precisely the squared absolute value of the quantum mechanical scattering amplitude Eq. (3.24).

We have assumed in the example just considered that there is only one point of stationary phase on the entire $\boldsymbol{b}$-plane. This means, in the classical limit, that only one particle trajectory corresponds to any value of the momentum transfer $\boldsymbol{q}$ and it is then no surprise that the differential cross section, when evaluated quantum mechanically, agrees closely with the classical result. When there are more trajectories that correspond classically to any value of $\boldsymbol{q}$, the quantum mechanical scattering amplitude will be a sum of terms f_j that introduce interference effects into the scattered intensity

$$f(\boldsymbol{k}', \boldsymbol{k}) = \sum_j f_j(\boldsymbol{k}', \boldsymbol{k}). \tag{3.28}$$

We shall show presently that these additional trajectories are supplied by further points of stationary phase in the plane of impact vectors $\boldsymbol{b}$, and that they are responsible for the presence of additional terms in the scattering amplitude, Eq. (3.28).

For the larger momentum transfers to which the asymptotic approximation applies most accurately it can provide dramatic insights into the behavior of scattering cross sections. It will frequently happen, for example, that Eq. (3.5) for the stationary phase point, or the equivalent Eq. (3.6) for the classical theory, possesses no solution for some range of momentum transfer vectors $\boldsymbol{q}$. In that case it is often possible to continue the phase function $\chi(\boldsymbol{b})$ analytically into a space of complex $\boldsymbol{b}$-vectors, to find stationary points in that space and then to carry out the necessary integrations by complex extensions of the same method we have illustrated. When applied in that way the method will furnish the quantum mechanical amplitudes for scattering processes that have lowered probabilities because they are classically forbidden.

It is also important, as we have noted in the Introduction, to be able to deal with intrinsically complex phase shift functions $\chi(\boldsymbol{b})$. Such phase shift functions also lead us inevitably to discuss complex values for the impact vectors $\boldsymbol{b}$. Indeed, they furnish a natural description of the absorptive effects brought about by inelastic scattering.

Our asymptotic technique for evaluating the scattering amplitude is not without limitations, however, and one is already evident in the calculations we have done. We have had to assume that the magnitude of $\boldsymbol{q}$ is large in order to justify using the method of stationary phase, and that limitation means we must not evaluate the scattering amplitude too close to the forward direction. The value of developing the asymptotic approximation further in order to correct it near the forward direction is indeed somewhat questionable since that is where it is easiest to evaluate the integral, Eq. (2.1), by more elementary and traditional means. It is also where the integral tends to present the least information about the spatial dependence of the interaction.

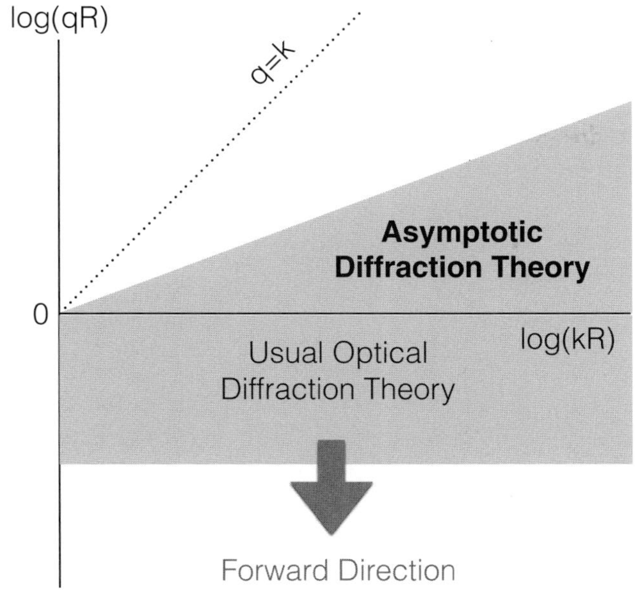

Figure 3.1 The region where the usual optical diffraction theory applies extends down to $\log(qR) \to -\infty$. The wedge $qR > 1$ contains the domain where the asymptotic theory is applicable. The size of the scattering target is represented by R, whereas the beam momentum and momentum transfer are denoted k and q, respectively.

To summarize in more mathematical terms, diffraction theory rests upon the two assumptions

$$q \ll k, \quad 1 \ll kR, \tag{3.29}$$

where R is a characteristic size of the scattering target, whereas the asymptotic limit also requires q to be large, $1 \ll qR$, or

$$\frac{1}{R} \ll q \ll k. \tag{3.30}$$

This regime, shown as a triangle in Fig. 3.1, is where our asymptotic considerations will apply.

4

Two Simple Examples

Having established a connection with classical scattering theory, we can now continue our introduction of the asymptotic method by discussing examples that are both elementary and useful. The first is the familiar case of the scattering of a charged particle by a Coulomb field and the second is the scattering by a much smoother interaction that is Gaussian in shape. The latter example will lead us to consider pairs of trajectories that contribute interfering amplitudes and we show in a third section how these paired trajectories can exhibit an interesting behavior that leads in the scattering of light by water droplets to the formation of rainbows.

4.1 Coulomb Scattering

The electrostatic potential energy of a charge e in the field of a point charge Ze at the origin ($r = 0$) is

$$V(\boldsymbol{r}) = \frac{Ze^2}{4\pi r}. \tag{4.1}$$

One of the familiar properties of this Coulomb potential is that its influence is felt so far from the origin that a wave incident upon it cannot, strictly speaking, be represented as a plane wave at any distance from the potential, however far away. The result, in the present context, is that the phase shift function $\chi(\boldsymbol{b})$ given by Eqs. (2.3) or (2.4) is logarithmically divergent. The problem presented by that behavior is removed however, by imagining that the field is electrostatically screened by an equal and opposite distribution of charge, for example, one uniformly distributed over a spherical shell, no matter at how great a distance from the origin. The screening can then be seen to have virtually no effect on the intensity of scattering at any but infinitesimal scattering angles.

We shall discuss screening in greater detail in Chapter 11, but let us first examine what is perhaps its most elementary example. The nuclear charge Ze is screened by a uniform spherical shell of radius R. The Coulomb potential is given by Eq. (4.1) for $r < R$, and it drops abruptly to zero for $r > R$. Then the phase shift function $\chi(\boldsymbol{b})$ is given by

$$\chi(b) = -\frac{2Ze^2}{4\pi\hbar v}\int_b^R \frac{dr}{\sqrt{r^2 - b^2}} \tag{4.2}$$

$$= -2\eta \log\left(\frac{R + \sqrt{R^2 - b^2}}{b}\right) \tag{4.3}$$

where η is the Sommerfeld parameter

$$\eta = \frac{Ze^2}{4\pi\hbar v}. \tag{4.4}$$

If the screening radius is of atomic dimensions, for example, it is so much larger than the wavelengths of typical nuclear projectiles that we may well consider R large enough to permit expanding the phase shift function as

$$\chi(b) = 2\eta \log\left(\frac{b}{2R}\right) + \mathcal{O}\left(\frac{b^2}{R^2}\right). \tag{4.5}$$

Retaining the terms of order b^2/R^2 and higher, in that case will tend only to influence the scattered intensity at angles θ of order $1/kR$ and smaller [9]. For scattering angles larger than this infinitesimal value, it suffices to write

$$\chi(b) = 2\eta \log\left(\frac{b}{2R}\right), \tag{4.6}$$

and it then becomes clear from Eq. (3.4) that the screening radius R only affects the scattering amplitude through an angle-independent phase factor.

Since $\chi(b)$ is rotationally symmetric for the Coulomb field, the stationary phase point determined by Eqs. (3.11b) and (3.12) has $b_y = 0$ and b_x determined by

$$\frac{2\eta}{b_x} = q. \tag{4.7}$$

It is worth remembering that q is an intrinsically positive variable, so that the stationary point

$$b_x = \frac{2\eta}{q}, \tag{4.8}$$

has the same sign as η, that is positive for repulsive forces and negative for attractive ones ($\eta \to -\eta$). For small angles of scattering the stationary point tends to lie far from the nucleus. As the scattering angle increases it draws closer to it.

The second derivative of the phase shift function at the stationary phase point is

$$\chi''(|b_x|) = -\frac{2\eta}{b_x^2} = -\frac{q^2}{2\eta}. \tag{4.9}$$

The scattering amplitude given by Eq. (3.23) is thus

$$f(\boldsymbol{k}', \boldsymbol{k}) = \frac{2\eta k}{q^2} \exp\left\{-2i\eta \log \frac{qR}{\eta} - 2i\eta + \frac{i\pi}{2}\right\} \tag{4.10}$$

and the differential cross section is thus given by the Rutherford formula

$$|f(\boldsymbol{k}', \boldsymbol{k})|^2 = \left(\frac{2\eta k}{q^2}\right)^2 = \frac{(2\eta k)^2}{|\boldsymbol{k} - \boldsymbol{k}'|^4}. \tag{4.11}$$

While this result could hardly be more familiar, the phase factor in the scattering amplitude, Eq. (4.10), may be less so. The precise form assumed for the shape of the screening function may easily be seen to alter the scattering amplitude by an angle-independent phase factor. We have taken a discontinuous model here to describe screening, but the result derived in Ref. [10] for exponential screening, for example,

$$\frac{2\eta k}{q^2} \exp\{-2i\eta \log qR + 2i \arg \Gamma(1 + i\eta)\}, \tag{4.12}$$

differs, after expanding $\arg \Gamma(1 + i\eta)$ to leading order in η, from the result of Eq. (4.10) only by the constant phase factor

$$\exp\{2i\eta \log \eta\}. \tag{4.13}$$

The above result in Eq. (4.11) for Coulomb scattering has been shown [14] to apply also to the case of gravitational interactions in small-angle scattering, where the effective strength however is energy-dependent.

4.2 Paired Trajectories

One of the many unusual features of scattering by the Coulomb field is that there is only one classical trajectory for each angle of scattering. For other sorts of interactions that number may be quite different. Sometimes two or more particle trajectories may contribute, occasionally there are even none that contribute in this way. To see a simple example of the latter sort of behavior let us consider a real-valued potential that decreases smoothly from some finite value to zero within a finite range. An example might be the Gaussian function $V(r) = V_0 \exp\{-r^2/R^2\}$. We shall discuss this example in detail in Chapters 7 and 11, but all we need for the moment is to let the phase shift function $\chi(\boldsymbol{b})$ be rotationally symmetric, and to take $\chi(b_x)$ to be an even-valued function with a similar shape, perhaps like that shown in Fig. 4.1(a).

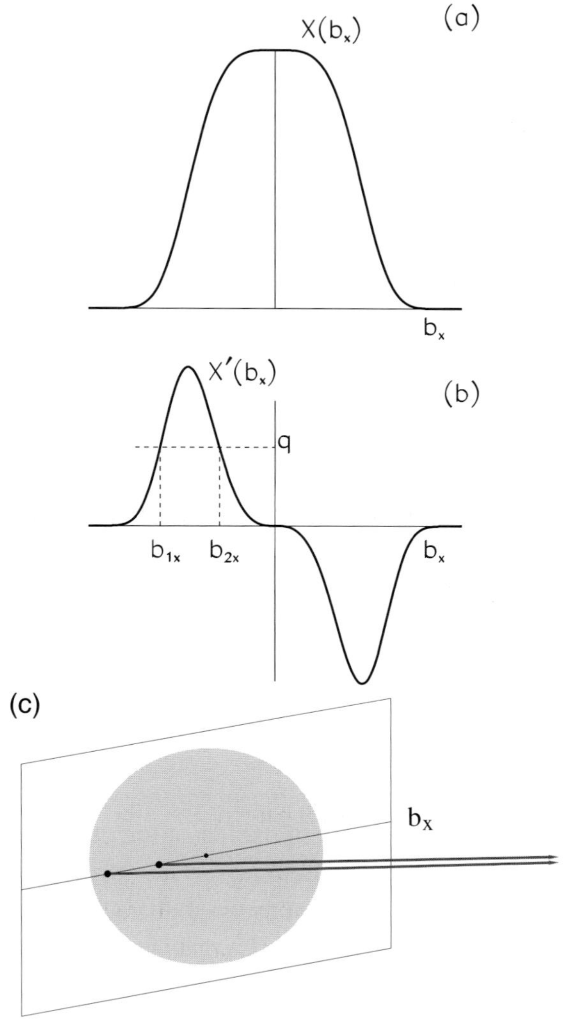

Figure 4.1 (a) A simple monotonic phase shift function, $\chi(b_x)$. (b) The equation $\chi'(b_x) = q$ has two solutions, b_{1x} and b_{2x}. (c) Schematic depiction of the trajectories corresponding to impact parameters b_{1x} and b_{2x}.

It follows then that $\chi'(b_x)$ is an odd function of b_x. The roots of the equation

$$\chi'(b_x) = q, \tag{4.14}$$

which is equivalent to Eq. (3.12), can immediately be seen from Fig. 4.1(b) to occur in pairs. For the case shown, which corresponds to an attractive potential, there are two stationary phase points at negative values of b_x, that are labelled b_{1x} and b_{2x}. The corresponding trajectories, shown in Fig. 4.1(c), both lead to the same momentum transfer $\hbar \boldsymbol{q}$.

If the scattering amplitudes that correspond to the two trajectories are $F(b_{1x})$ and $F(b_{2x})$ respectively, then the classical differential cross section is the sum of the two corresponding intensities,

$$\left(\frac{\mathrm{d}\sigma}{\mathrm{d}\Omega}\right)_{\text{class.}} = |F(b_{1x})|^2 + |F(b_{2x})|^2. \tag{4.15}$$

The quantum mechanical differential cross section, on the other hand, will be given, as long as the two roots b_{1x} and b_{2x} remain distinct, by the intensity

$$\frac{\mathrm{d}\sigma}{\mathrm{d}\Omega} = |f(\boldsymbol{k}', \boldsymbol{k})|^2 = |F(b_{1x}) + F(b_{2x})|^2 \tag{4.16}$$

that results from superposing the two amplitudes before squaring. Because there is no way of knowing which of the two trajectories the particle has followed the cross section will contain interference terms, and these, as we shall see, go a long way toward explaining the oscillations present in cross sections like those of Fig. 1.1.

4.3 Rainbows

It is clear from Fig. 4.1(a) that as we increase the value of q the two roots b_{1x} and b_{2x} will approach each other and for a certain value of q will coalesce. For larger values of q there are no roots, which is to say there is no classical trajectory at all leading to that momentum transfer. This disappearance of the classical trajectories or "rays" is quite familiar in the theory of the formation of the rainbow. The principal difference is that for scattering by raindrops it takes place at a rather backward angle of about 139°, rather than outside a forward diffraction cone.

We may pause for a moment to note that this element of the theory of the rainbow was one of the most remarkable insights of René Descartes. Descartes seems to have discovered the law of refraction, perhaps independently of Snell, during a lengthy stay in Holland. He realized that the angular distribution of light rays refracted and reflected by spherical raindrops would be independent of their radii, which he knew to vary considerably. He found the angles of refraction and reflection of the variously impinging rays, and showed that for the combination of two refractions and one internal reflection, the rays have a minimum angle of deflection (or scattering) of about 139°*, relative to the forward direction. There are two possible ray paths that lead to scattering through angles still larger than that. What is perhaps most impressive was Descartes' realization, in 1636, long before the invention of calculus, that the extremal character of this* 139° *deflection would lead to a concentration of the scattered light there. He had indeed explained most of the geometry of the rainbow and, by including the possibility of a second internal reflection, was even able to explain the corresponding features of the secondary*

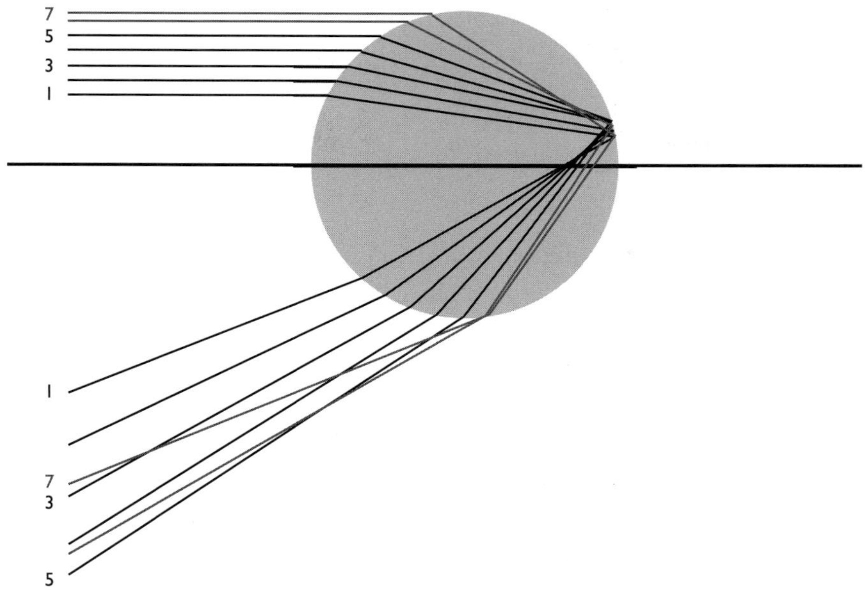

Figure 4.2 Parallel rays of light entering from top left and undergoing rainbow scattering. Only every second one is labeled for lack of space. The rays 6 and 7 hit at an impact parameter beyond the rainbow point and get scattered to angles inside the rainbow.

rainbow. But because the discovery of dispersion by Newton was still many years off, Descartes' rainbows lacked altogether any spectral separation of colors which is their principal hallmark. They were simply bright arcs of light just as white as direct sunlight.

Let us write q_R for the extremal value of q, which defines the "rainbow" angle of deflection for waves of a particular wavelength λ and let $b_{x,R}$ be the value at which the roots b_{1x} and b_{2x} coalesce. It is quite instructive to discuss the behavior of the asymptotic approximation for q-values in the neighborhood of q_R. We can assume that in the immediate neighborhood of its maximum the function $\chi'(b_x)$, as shown in Fig. 4.3, takes the form

$$\chi'(b_x) \simeq q_R - C(b_x - b_{x,R})^2 \tag{4.17}$$

where C is some positive constant. Then for $q < q_R$ the roots of Eq. (4.14) are given by

$$b_{1x} = b_{x,R} - \sqrt{\frac{q_R - q}{C}} \tag{4.18a}$$

$$b_{2x} = b_{x,R} + \sqrt{\frac{q_R - q}{C}}, \tag{4.18b}$$

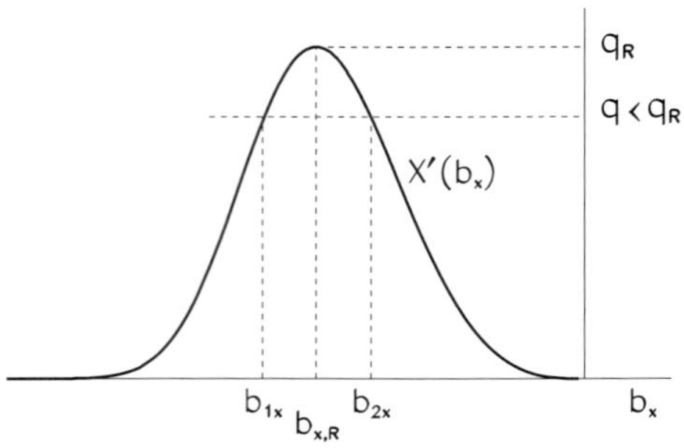

Figure 4.3 The occurrence of a rainbow at the maximum of the deflection function $\chi'(b_x)$.

and the corresponding values of χ'' are

$$\chi''(b_{1x}) = 2\sqrt{C(q_R - q)} \tag{4.19a}$$

$$\chi''(b_{2x}) = -2\sqrt{C(q_R - q)}. \tag{4.19b}$$

The two stationary points contribute, according to Eq. (3.23), the scattering amplitudes

$$F_1(b_{1x}) = k\left\{\frac{b_{1x}}{2q\sqrt{C(q_R - q)}}\right\}^{\frac{1}{2}} e^{i\{-qb_{1x}+\chi(b_{1x})\}}, \tag{4.20a}$$

$$F_2(b_{2x}) = k\left\{\frac{-b_{2x}}{2q\sqrt{C(q_R - q)}}\right\}^{\frac{1}{2}} e^{i\{-qb_{2x}+\chi(b_{2x})\}}. \tag{4.20b}$$

Both of these amplitudes become singular as $(q_R - q)^{-\frac{1}{4}}$ as q approaches the "rainbow" value q_R, and that indeed is the essence of Descartes' early observation. In the classical limit of zero wavelength, which corresponds to the limit of ray optics, the scattered intensity rises as $(q_R - q)^{-\frac{1}{2}}$ as q approaches q_R and then drops abruptly to zero. There is no scattering for larger momentum transfers, no illumination outside the rainbow cone.

In quantum mechanics, however, as in wave optics, the picture is somewhat different. The intensity may reach a strong maximum at the rainbow point $q = q_R$, but it is not truly singular there. Furthermore scattering to larger momentum transfers is not completely forbidden. It can indeed take place with easily observed intensities as a kind of barrier penetration phenomenon.

We shall discuss in Chapter 7 a simple way of correcting the singular behavior of scattering amplitudes like Eq. (4.20). These singularities are so localized in their occurrence and their effects, however, that the problem of regularizing them is not

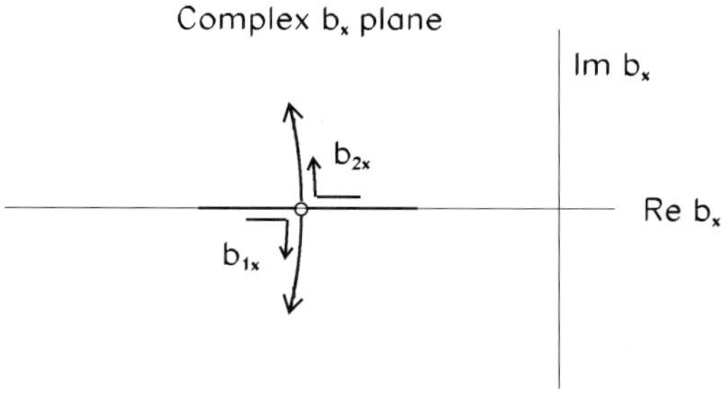

Figure 4.4 The impact parameters b_{1x} and b_{2x} in the region of the rainbow point $b_{x,\mathrm{R}}$ (denoted by the circle). For $q > q_\mathrm{R}$ they become complex.

often an important one. What is far more important is to gain some understanding of the non-classical scattering that is present beyond the rainbow angle. To do that let us begin by noting that for $q > q_\mathrm{R}$ the stationary points do not disappear altogether. They simply move off the real axis into the complex plane of b_x-values. For $q > q_\mathrm{R}$ we can thus write

$$b_{1x} = b_{x,\mathrm{R}} - i\sqrt{\frac{q - q_\mathrm{R}}{C}} \tag{4.21a}$$

$$b_{2x} = b_{x,\mathrm{R}} + i\sqrt{\frac{q - q_\mathrm{R}}{C}}. \tag{4.21b}$$

This shift of the roots into the complex plane of b_x is illustrated in Fig. 4.4.

For values of q inside the rainbow angle, $q < q_\mathrm{R}$, the integration over all values of b_x required in Eq. (3.4) actually passes through two stationary points, b_{1x} and b_{2x}. When q exceeds q_R, those points are complex, and the path of integration of b_x, along the real axis from $-\infty$ to ∞ passes through neither of them. We can easily shift the path of integration into the complex plane by adding an imaginary constant to b_x. That could pass the path through one of the stationary points, but which one should we choose?

It is important to note at this stage that the second derivative of the phase function $\chi(b_x)$ for $q > q_\mathrm{R}$ takes the values

$$\chi''(b_{1x}) = 2i\sqrt{C(q - q_\mathrm{R})} \tag{4.22a}$$

$$\chi''(b_{2x}) = -2i\sqrt{C(q - q_\mathrm{R})} \tag{4.22b}$$

at the stationary points, while $1/b_x$ (cf. Eq. (3.18a)) takes the values

$$\frac{1}{b_{1x}} = \frac{1}{b_{x,\mathrm{R}} - i\beta(q)} \tag{4.23a}$$

$$\frac{1}{b_{2x}} = \frac{1}{b_{x,\mathrm{R}} + i\beta(q)} \tag{4.23b}$$

where

$$\beta(q) \equiv \sqrt{\frac{q - q_{\mathrm{R}}}{C}}. \tag{4.24}$$

For the roots we have designated as b_{1x}, both $1/b_{1x}$ and χ'' have positive imaginary parts. By referring back to Eqs. (3.18a), (3.18b), (3.21a), and (3.21b) we see that the b_x- and b_y-integrations, when carried out through this point will both converge as the integral of a Gaussian function, which is to say quite rapidly. Integrating through the point b_{2x} on the other hand would furnish divergent results for both. The most natural procedure therefore is to shift the integration path in the negative imaginary direction of the b_x-plane by introducing the variable

$$b'_x = b_x - i\beta(q) \tag{4.25}$$

and then integrating over real values of b'_x from $-\infty$ to ∞. That procedure, it is easy to show, leads to no change in the value of the integral Eq. (3.21b) as long as the imaginary part of the parameter we have called α_x is positive.

In this way we see that for $q > q_{\mathrm{R}}$ only the stationary point b_{1x} contributes to the scattering. It furnishes, according to Eq. (3.23), the scattering amplitude

$$F(b_{1x}) = k\left\{\frac{b_{1x}}{2iCq\beta(q)}\right\}^{\frac{1}{2}} \exp\{-q\beta(q) - iqb_{x,\mathrm{R}} + i\chi(b_{1x})\}. \tag{4.26}$$

Certain qualitative features of the differential cross section this expression provides are worth noting immediately. First of all, the interference effects present when two or more stationary points contribute are absent in this case, so the differential cross section may be expected to show less oscillation outside the rainbow cone. Secondly, if the stationary point b_{1x} were fixed at some value and its slow dependence on q neglected, the most important remaining dependence on q would be $F \sim \exp\{-q\beta\}$. This is a result reminiscent of the single-slit model we discussed in the introduction, cf. Eq. (1.5), and we have thus made a first step towards it. In this sense the displacement of our stationary phase point from the real axis yields an amplitude that corresponds to our earlier model of a slit of width β. The accuracy of this correspondence depends, of course, on the degree to which the stationary point moves as q varies. In the examples of interactions of finite range that we shall consider presently, the stationary point usually shifts less rapidly for large q than it does in the present simplified example. The result is that the square root factor and the factor $\exp\{i\chi(b_{1x})\}$ in the expression in Eq. (4.24) vary much more slowly than the factor $\exp\{-q(\beta + ib_{x,\mathrm{R}})\}$, which, as we have noted, could be obtained from a model of diffraction by a single slit, one that is slightly refracting and has diffusely absorbing edges.

5

Complex Trajectories

There are excellent reasons, as we have seen, to consider the possibility of complex stationary points, even when the particle interactions are purely real. When we account for the influence of inelastic transitions by regarding them as absorption and thereby making the phase shift function $\chi(\boldsymbol{b})$ complex, the more general consideration of complex trajectories becomes inevitable. It is therefore worth reconsidering, for a moment, some of the calculations we have carried out in Chapter 3 in terms of real variables.

In these situations when classical trajectories are absent, there remains a quantum mechanical scattering amplitude which can be represented by venturing into the complex plane. Let us again assume that the phase shift function $\chi(\boldsymbol{b})$ has rotational symmetry and so depends only on $|\boldsymbol{b}| = b$. Then the stationary phase point will still have $b_y = 0$, according to Eq. (3.11b). We must deal with the fact that the roots of Eq. (3.11a) for b_x, however, may lie off the real axis in the complex b_x-plane. But how do we even define $\chi(b)$ in general for complex values of b_x? At any stationary point we have $b_y = 0$, and therefore we need only discuss the phase function $\chi(b_x, 0)$, which is always an even function of b_x since $b = \sqrt{b_x^2 + b_y^2}$,

$$\chi(-b_x, 0) = \chi(b_x, 0), \tag{5.1}$$

(while $\chi(b)$ need not be an even function of b, regarded as a complex variable). The condition to determine the b_x components of the stationary phase points, Eq. (3.8), is then still

$$\frac{\partial}{\partial b_x}\chi(b_x, 0) = q. \tag{5.2}$$

We shall assume that both the real and imaginary parts of $\chi(b_x, 0)$ are analytic for values of b_x along the real axis, i.e., we can write

$$\chi(b_x, 0) = \mathcal{R}(b_x) + i\mathcal{I}(b_x), \tag{5.3}$$

where $\mathcal{R}(b_x)$ and $\mathcal{I}(b_x)$ are real-valued analytic functions for real b_x. Then both of these functions can be continued analytically into the complex b_x-plane, where they will then obey the reflection principle

$$\mathcal{R}(b_x^*) = \mathcal{R}^*(b_x), \qquad \mathcal{I}(b_x^*) = \mathcal{I}^*(b_x). \tag{5.4}$$

The phase shift $\chi(b_x, 0)$ as a whole, however, since it is complex, will *not* obey the reflection principle, i.e.,

$$\chi^*(b_x, 0) = \mathcal{R}^*(b_x) - i\mathcal{I}^*(b_x) \neq \chi(b_x^*, 0). \tag{5.5}$$

The symmetries in Eq. (5.4) are nonetheless useful in locating the roots of Eq. (5.2).

For purely real phase shifts χ, for example, we can take the complex conjugate of the equation to determine the stationary point,

$$\frac{\partial}{\partial b_x}\mathcal{R}(b_x) = q \tag{5.6}$$

by writing

$$\frac{\partial}{\partial b_x^*}\mathcal{R}^*(b_x) = \frac{\partial}{\partial b_x^*}\mathcal{R}(b_x^*) = q. \tag{5.7}$$

If b_x is a root of Eq. (5.6) then so is b_x^*. For real interactions then the roots can come in complex conjugate pairs, much as we found in the examples of Section 4.2.

For purely imaginary phase shifts, on the other hand, a root must satisfy the equation

$$i\frac{\partial}{\partial b_x}\mathcal{I}(b_x) = q. \tag{5.8}$$

If we take the complex conjugate of this equation and avail ourselves of the symmetry conditions in Eqs. (5.1) and (5.4), we see that

$$-i\frac{\partial}{\partial b_x^*}\mathcal{I}^*(b_x) = -i\frac{\partial}{\partial b_x^*}\mathcal{I}(b_x^*) = -i\frac{\partial}{\partial b_x^*}\mathcal{I}(-b_x^*) = q. \tag{5.9}$$

Hence for each b_x that satisfies Eq. (5.8) there will be another root $-b_x^*$, which is its reflection in the imaginary axis.

To proceed with the evaluation of the scattering amplitudes we shall assume again as we did in Chapter 3 that the function $\chi(\boldsymbol{b}) - \boldsymbol{q}\cdot\boldsymbol{b}$ can be expanded quadratically about its stationary points $\boldsymbol{b}_1$ and $\boldsymbol{b}_2$. We can then use the same summation convention we used in Chapter 3, i.e., in connection with Eqs. (3.20) and (3.21) to define the required integrals. The integration over b_y still has a stationary point at $b_y = 0$ and essentially takes the form of Eq. (3.21a). It can still be carried out over the b_y axis from $-\infty$ to ∞, but its convergence requires that the imaginary part of α_y, defined by Eqs. (3.17) and (3.18b) not be negative. If $\boldsymbol{b}_j$ is the j-th stationary phase point then, to carry out the b_y integration for this point we require, according to Eq. (3.18b)

$$\operatorname{Im}\alpha_y = \frac{q}{2}\operatorname{Im}\left(\frac{1}{b_{jx}}\right) \geq 0. \tag{5.10}$$

This important constraint shows that the b_y integrals are convergent only for stationary points b_{jx} lying in the lower half of the b_x-plane.

The integration over b_x, however, has two stationary points at which the modulus of the integrand has maxima. If the stationary points are sufficiently far apart, or the maxima at these points are narrow enough, we can approximate that integral as the sum of two Gaussian integrals similar to Eq. (3.21b).

The case of two or more stationary points is best discussed for the absorptive case. Then the argument of the exponential can be expanded to second order in $b_x - b_{jx}$ around each stationary point, with a coefficient (earlier denoted as α_x) that has a positive imaginary part. It is that positive imaginary part which provides convergence. When this coefficient is sufficiently large, the contributions to the integral over b_x will be very localized, and there exists a point b_{0x} somewhere between the two stationary points b_{1x} and b_{2x}, such that it is meaningful to split the range of integration in Eqs. (3.20) and (3.21b) into two:

$$\begin{aligned}
&\lim_{\epsilon\to 0}\int\int e^{-\epsilon(b_x^2+b_y^2)-iqb_x+i\alpha_y b_y^2+i\chi(|b_x|)}\, db_x\, db_y \\
&\simeq \lim_{\epsilon\to 0}\int_{-\infty}^{\infty}\sqrt{\frac{i\pi}{\alpha_y}}\, e^{-\epsilon b_x^2-iqb_x+i\chi(|b_x|)}\, db_x \\
&= \lim_{\epsilon\to 0}\left[\int_{-\infty}^{b_{0x}} + \int_{b_{0x}}^{\infty}\right]\sqrt{\frac{i\pi}{\alpha_y}}\, e^{-\epsilon b_x^2-iqb_x+i\chi(|b_x|)}\, db_x \\
&\simeq e^{-iqb_{1x}+i\chi(|b_{1x}|)}\sqrt{\frac{i\pi}{\alpha_{1y}}}\int_{-\infty}^{b_{x0}} e^{i\alpha_{1x}(b_x-b_{1x})^2}\, db_x \\
&\quad + e^{-iqb_{2x}+i\chi(|b_{2x}|)}\sqrt{\frac{i\pi}{\alpha_{2y}}}\int_{b_{x0}}^{\infty} e^{i\alpha_{2x}(b_x-b_{2x})^2}\, db_x .
\end{aligned} \tag{5.11}$$

Here, α_{1x} and α_{2x} refer to the second derivatives of the phase shift function at the points of stationary phase,

$$\alpha_{1x} \equiv \tfrac{1}{2}\chi''(|b_{1x}|), \quad \alpha_{2x} \equiv \tfrac{1}{2}\chi''(|b_{2x}|). \tag{5.12}$$

Likewise, α_{1y} and α_{2y} refer to the second derivatives of the phase shift function with respect to b_y, as discussed in Chapter 3, and evaluated at the stationary points b_{1x} and b_{2x}. Furthermore, the assumed convergence will allow us to extend the first integral up to $b_x = \infty$ and the second one down to $b_x = -\infty$, leading to two distinct contributions, each of the form given in Eq. (3.21b).

5.1 A Purely Absorptive Interaction

When short-range inelastic interactions such as particle production processes completely dominate the elastic ones, as they do for collisions at sufficiently high energies, it has often been found useful to regard the overall interaction as purely imaginary or absorptive.

As an analytically simple example of an absorptive interaction, let us assume that the phase shift function takes the form

$$\chi(b) = \frac{iA}{\cosh(b/\beta)}, \tag{5.13}$$

where β is the range of the interaction and A is a positive-valued opacity parameter. We shall see this example as a useful one to consider in detail as a model for the calculations to be undertaken later. Our choice of an even function of b, for example, means that $\chi(b_x, 0)$ is a single-valued function of b_x,

$$\chi(b_x, 0) = \frac{iA}{\cosh(b_x/\beta)}. \tag{5.14}$$

It is furthermore periodic under imaginary displacements of b_x with period $2\pi\beta$.

The stationary points are the roots of Eq. (5.2), that is to say the roots of

$$-\frac{iA}{\beta}\frac{\sinh(b_x/\beta)}{\cosh^2(b_x/\beta)} = q. \tag{5.15}$$

If we introduce the parameter

$$\lambda = \frac{A}{2q\beta}, \quad (\lambda > 0), \tag{5.16}$$

then Eq. (5.15) can be written as the quadratic equation

$$\sinh^2(b_x/\beta) + 2i\lambda \sinh(b_x/\beta) + 1 = 0, \tag{5.17}$$

which has the solutions

$$\sinh(b_x/\beta) = i\{-\lambda \pm \sqrt{1+\lambda^2}\}. \tag{5.18}$$

To find the stationary phase points b_x we can let

$$\frac{b_x}{\beta} = x + iy.$$

Then we find

$$\sinh(b_x/\beta) = \sinh x \cos y + i \cosh x \sin y, \tag{5.19}$$

and we see from Eq. (5.18) that the real term in this expression must vanish,

$$\sinh x \cos y = 0. \tag{5.20}$$

Hence either $\cos y$ or $\sinh x$ must vanish. If $\cos y = 0$ we must have

$$y = (n - \tfrac{1}{2})\pi \tag{5.21}$$

for any integer value of n, and thus

$$\cosh x = (-1)^{n+1}\big(-\lambda \pm \sqrt{1+\lambda^2}\big). \tag{5.22}$$

Since λ is intrinsically positive, and we require x to be real, this equation will only have solutions for even values of n. In those cases the roots are given by

$$x = \pm \cosh^{-1}\big(\lambda + \sqrt{1+\lambda^2}\big), \tag{5.23a}$$
$$y = (2m - \tfrac{1}{2})\pi \quad m = -2, -1, 0, 1, 2, \ldots \tag{5.23b}$$

If, on the other hand, it is $\sinh x$ that vanishes, it follows that $x = 0$ and in that case y must satisfy

$$\sin y = -\lambda \pm \sqrt{1+\lambda^2}. \tag{5.24}$$

The only solutions for real y are those for which $\sin y$ takes the positive value, or those for which we have the pair of values

$$x = 0, \tag{5.25a}$$
$$y = \left(2m - \tfrac{3}{2}\right)\pi \pm \cos^{-1}\big(\sqrt{1+\lambda^2} - \lambda\big), \quad m = -2, -1, 0, 1, 2, \ldots \tag{5.25b}$$

Both the sets of roots in Eqs. (5.23) and (5.25) possess the required reflection symmetry in the imaginary axis.

For small values of λ, or large values of the momentum transfer, the roots of Eq. (5.15) are quite naturally found to be clustered in pairs, one pair near each of the poles of $\chi(b)$. (See Fig. 5.1.) For large λ on the other hand, or small momentum transfers, the pairs of roots that are off the imaginary axis tend to be spread far apart in their x coordinates. Those that are on the imaginary axis tend to approach the values $y = 2\pi n i$ where n is an integer. (See Fig. 5.2.)

We have found the stationary points, in fact an infinite number of them, and so we turn next to the problem of how they are to be used in approximating the diffraction integral, Eq. (3.4). Let us recall, first of all, that convergence of the integral over b_y has put a strong restriction on the values of b_x. We see in Eq. (5.10) that at the

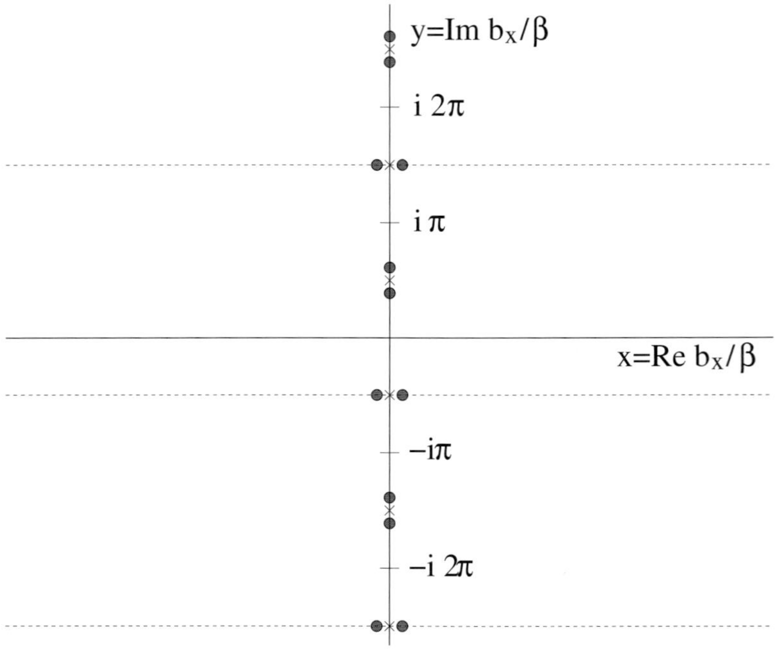

Figure 5.1 Solid circles are the roots of Eq. (5.15) for large momentum transfers, or $\lambda = A/2q\beta \ll 1$. Note that they are clustered in pairs about the poles (crosses) of the function $\chi(b)$ given by Eq. (5.13).

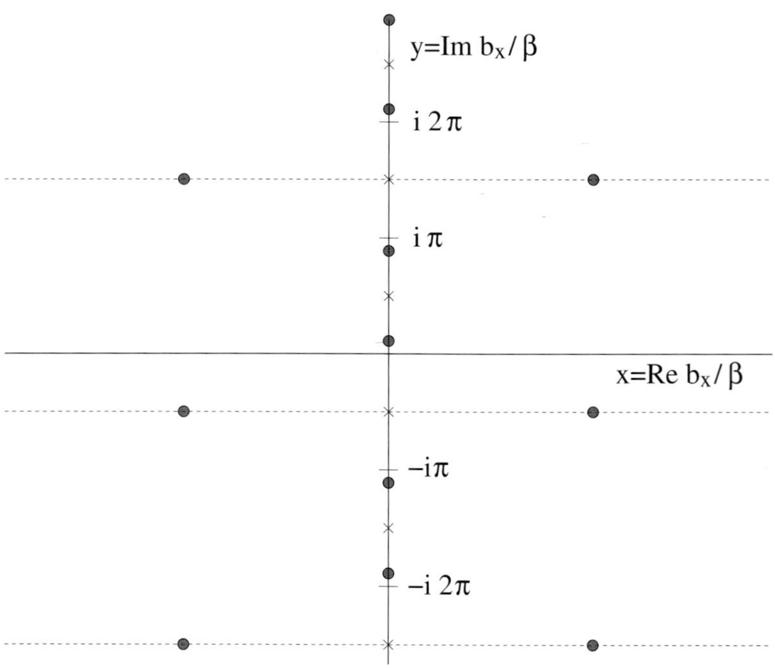

Figure 5.2 Solid circles are the roots of Eq. (5.15) for small momentum transfers or $\lambda = A/2q\beta \gg 1$.

stationary points which lie on the path of integration, b_x must not lie in the upper half-plane. The requirement

$$\mathrm{Im}\, b_{jx} \leq 0 \tag{5.26}$$

eliminates half the roots we have found, but still leaves an infinite number of candidates for possible use.

5.2 Choice of Integration Path

The integral that we seek, according to Eq. (3.4), is to be carried out over the real b_x-axis. Given the analytic properties we have assumed for $\chi(b_x, 0)$, it will be possible in general to deform the path of integration into the lower half-plane without changing the value of the integral. That will be true as long as $\chi(b_x, 0)$ is analytic in the region between the new path and the real axis. If we are to use the approximation methods illustrated in Chapter 4, we must steer the integration path through stationary phase points in such a way that the dominant contribution to the integral come just from those points. That may seem, at first, a strange assignment, but we have no need, of course, to use all of the stationary points that are available, and in the present case two will do.

Once the b_y integration has been carried out in Eq. (3.4) the function we must integrate over b_x is

$$h(b_x) = \exp\{-\epsilon b_x^2 - iqb_x + i\chi(b_x, 0)\}. \tag{5.27}$$

To determine an appropriate contour of integration we note that $\chi(b_x, 0)$ goes exponentially to zero as $\mathrm{Re}\, b_x \to \pm\infty$. Then $h(b_x)$ goes strongly to zero as well for $|\,\mathrm{Re}\, b_x| \to \infty$, thanks to the ϵb_x^2 term in the exponent. The integral of $h(b_x)$ then, taken over any finite segment of path, will vanish in the limit in which the segment lies at distances for which $|\,\mathrm{Re}\, b_x| \to \infty$.

If we deform the path, for example, to the one shown in Fig. 5.3, the two vertical segments will give vanishing contributions in the limit as they recede to infinity,

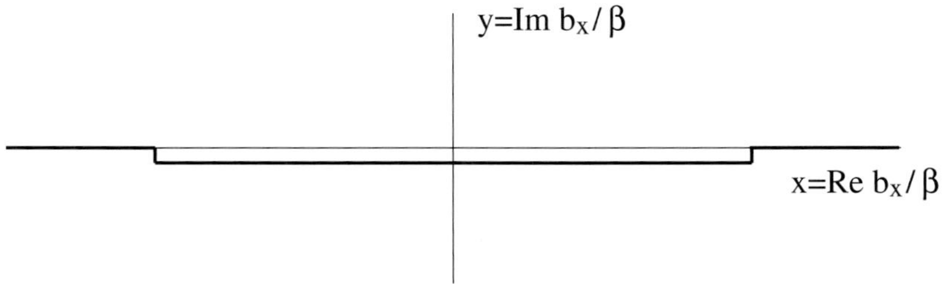

Figure 5.3 Displacement of the path of integration away from the real b_x-axis.

and the integral over the horizontal segment will be the same as the integral over the real axis—provided $h(b_x)$ is analytic in the strip between them. So we are free to displace the straight-line path downwards, for example, but we must pay attention to the poles of $\chi(b_x, 0)$ along the imaginary axis, which are essential singularities of $h(b_x)$.

Let us translate the straight-line path into the lower half-plane until it passes through the two stationary points labelled by $m = 0$ in Eq. (5.23). That is as far as we can translate it, in fact, without passing the singularity at $b_x = -i\frac{\pi}{2}\beta$. Along this line, that is to say for $y = -\frac{\pi}{2}$, we have

$$\chi(b_x, 0) = -\frac{A}{\sinh x}, \tag{5.28}$$

and if we let $\epsilon \to 0$, we can write

$$h(b_x) = \exp\left\{-\frac{\pi q\beta}{2} - i\left[q\beta x + \frac{A}{\sinh x}\right]\right\}. \tag{5.29}$$

The modulus of the function h is constant along this line. Only its phase varies. The variation of the phase along the line is shown in Fig. 5.4. It exhibits explicitly the two stationary points given by Eq. (5.23a). The straight path $b_x = \beta(x - \frac{i\pi}{2})$, in other words, represents just the sort of contour we want for the stationary phase approximation.

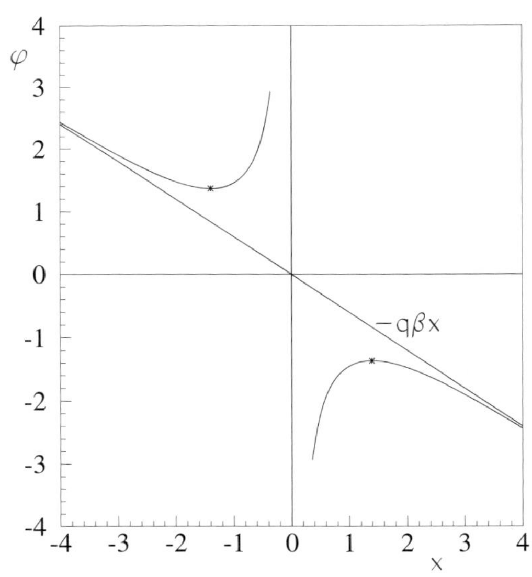

Figure 5.4 Variation of the phase $\varphi = \arg h(b_x)$ of Eq. (5.29) along the line $b_x = \beta(x - i\pi/2)$. The stationary phase points indicated by $*$ are given by Eq. (5.23a).

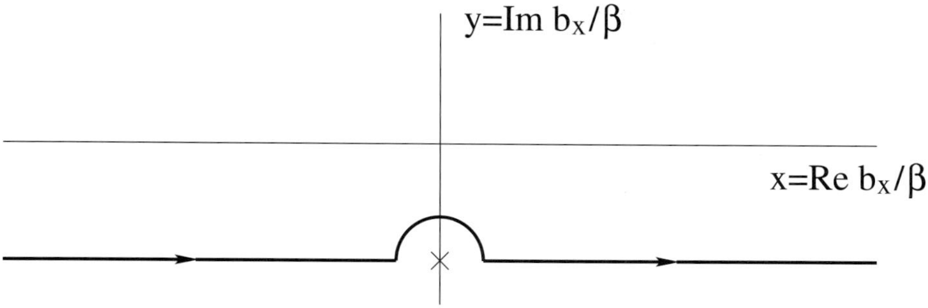

Figure 5.5 Deformation of the straight-line contour to avoid the singularity at $b_x = -i\pi/2$.

It should be noted that in picking this contour we are choosing to integrate through an essential singularity of $h(b_x)$ at $x = 0$, $y = -\frac{\pi}{2}$. Mathematically sensitive souls may regard that as risky if not absolutely wicked, so a few more words may be in order. We can easily deform the contour to follow an infinitesimal semicircle above the singularity and thereby avoid it altogether. (See Fig. 5.5.) On this semicircle we can write

$$b_x = \beta\left(-\frac{i\pi}{2} + \delta e^{i\theta}\right) \tag{5.30}$$

for an infinitesimal value of the radius δ, and any phase angle θ. The phase function $\chi(b_x, 0)$ then has an imaginary part

$$\mathrm{Im}\,\chi(b_x, 0) \sim \frac{A \sin\theta}{\delta} \tag{5.31}$$

which becomes positively infinite for angles θ such that $0 < \theta < \pi$. The modulus of $h(b_x)$ thus vanishes on the semicircle, and that justifies the more brutal procedure of integrating straight through the singularity.

There is, of course, an infinite variety of contours that all yield the same integral for an analytic function. The question is simply how suitable they are for various methods of approximation. Could we have chosen the integration path differently for our example? The answer is obviously "yes". We have made no mention thus far of any more general properties of the stationary phase points defined by Eq. (5.2). They are points at which the derivative of $h(b_x)$ vanishes (for $\epsilon = 0$) in the complex analytic sense, and are therefore saddle points in the modulus of $h(b_x)$. A contour plot of the function $\mathrm{Im}\,\{qb_x - \chi(b_x, 0)\}$ which determines the contours of $|h(b_x)|$ is shown in Fig. 5.6 for $q\beta = 1.2$. The stationary phase points are seen to be saddle points as well. Instead of using the method of stationary phase to approximate the integral of $h(b_x)$, we could, with equal justice, have used the closely related method of steepest descents [17]. Had we done that, we should formally have chosen the

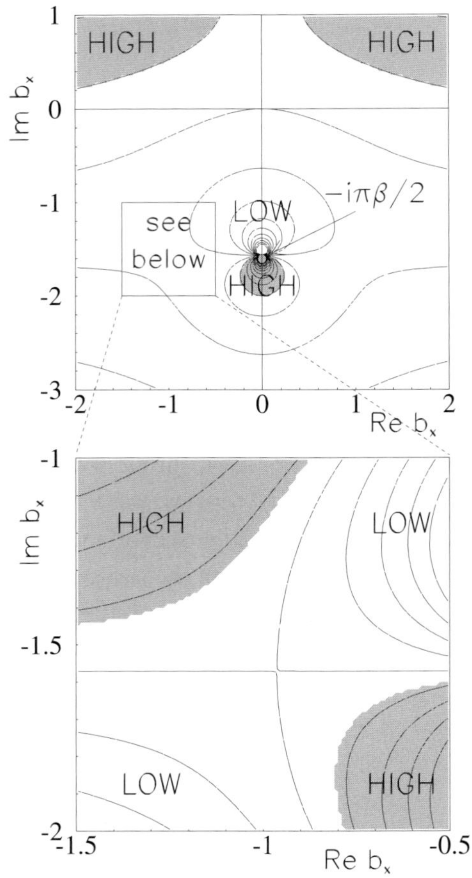

Figure 5.6 Contour plots of the function $\psi = \text{Im}\{qb_x - \chi(b_x, 0)\}$. In the upper part a larger area is shown, including the singularity at $b_x = -i\pi\beta/2$. Shaded: $\psi > 0$. In the lower part we show details of the saddle point to the left of the singularity (shaded: $\psi > -1.8$). There is a similar saddle point to the right of the singularity, for $\text{Re}\, b_x > 0$.

contour that comes in from $b = -\infty - i\infty$ along the low "valley" in the lower left of Fig. 5.6 (note that there is no contribution from $b_x = -\infty$ to $b_x = -\infty - i\infty$), goes over the saddle point, into the deep region between the singularity on the imaginary axis and the origin, and out again in the "valley" (not shown) towards $b_x = \infty - i\infty$. But in practice we should have approximated the integrals through the saddle points by integrations over straight lines tilted at $\pm 45°$ with respect to the real axis.

The results then would be precisely the same as those we find by using the method of stationary phase and the more convenient straight-line contour.

We have already made implicit use of this alternative method, in fact, in our discussion of rainbow behavior in Chapter 4. There the integration of a function that is oscillating rapidly for $q < q_R$ turned magically into an integration over decreasing Gaussian functions for $q > q_R$. The method of stationary phase had in effect become the method of steepest descents because of the appearance of a factor of i in the coefficient α_x of Eq. (3.18a).

As long as we transit the stationary points along straight paths, and use quadratic approximations to the behavior of $\chi(b_x, 0)$ near those points, the two methods are effectively the same. We can even make a continuum of compromises between them without altering the results. There will always be a 90° interval of angles within which the path of integration can transit the stationary point so that the point represents a maximum of the modulus rather than a minimum. We shall often find it convenient to use such intermediate paths when the behavior of the function $\chi(b_x, 0)$ is less symmetric in the complex plane.

The technique of approximating the integral along the line $\operatorname{Im} b_x = -\pi/2$ is no different from that we have used before for integrating along the real axis. There are two stationary points on the line, which we label b_{1x} and b_{2x} respectively. These are given by

$$\operatorname{Re} b_{1x} = -\operatorname{Re} b_{2x} = \beta \cosh^{-1}\left(\lambda + \sqrt{1+\lambda^2}\right) \tag{5.32a}$$

$$\operatorname{Im} b_{1x} = \operatorname{Im} b_{2x} = -\pi\beta/2 \tag{5.32b}$$

with λ defined by Eq. (5.16). The integral in Eq. (3.4) is then approximated as a sum of two terms, one contributed by each of the stationary points. These partial scattering amplitudes are given, as in Eq. (3.23), by

$$f_j(\boldsymbol{k}', \boldsymbol{k}) = \frac{k}{i}\left\{\frac{-b_{jx}}{q\chi''(b_{jx}, 0)}\right\}^{\frac{1}{2}} e^{-iqb_{jx} + i\chi(b_{jx}, 0)}, \quad (j = 1, 2). \tag{5.33}$$

The overall scattering amplitude is then

$$f(\boldsymbol{k}', \boldsymbol{k}) = f_1(\boldsymbol{k}', \boldsymbol{k}) + f_2(\boldsymbol{k}', \boldsymbol{k}). \tag{5.34}$$

The scattering amplitudes f_j possess a symmetry worth noting. Since $b_{2x} = -b_{1x}^*$, we also have, for the phase shift function in Eq. (5.13),

$$\chi(b_{2x}, 0) = -\chi^*(b_{1x}, 0) \tag{5.35}$$

$$\chi''(b_{2x}, 0) = -\chi''^*(b_{1x}, 0). \tag{5.36}$$

It follows then that

$$f_2(\boldsymbol{k}', \boldsymbol{k}) = -f_1^*(\boldsymbol{k}', \boldsymbol{k}), \tag{5.37}$$

and that the total scattering amplitude, Eq. (5.34), is purely imaginary, as it must be for a purely absorptive interaction.

We shall not undertake the numerical evaluation of the scattering amplitude in Eq. (3.23) here, but will save it for Chapter 8, where we begin the analysis of a number of analogously shaped interactions, and include the possibility of refraction as well as absorption. We shall show there that for appropriate ranges of parameters our approximate scattering amplitude adheres quite closely in numerical terms to the behavior of the general expression for the diffraction amplitude, Eq. (2.1), from which we began.

Even before substituting any numbers however, we can see certain general features of the scattering described by Eq. (3.23). The dominant dependence on q in the amplitude f_j will lie in the factors

$$\exp(-iqb_{jx}) = \exp\left\{-iq\,\mathrm{Re}\,b_{jx} - \frac{\pi q\beta}{2}\right\}. \tag{5.38}$$

Both of the amplitudes for $j = 1$ and 2 will decrease in modulus exponentially with increasing q. Each will correspond then approximately to the amplitude transmitted by a small diffuse aperture of the kind we have discussed in Chapter 1. The diffuseness parameter we called β there is replaced by $\pi\beta/2$ here.

The phase factors contained in the exponentials in Eq. (5.38) also depend strongly on q. When the sum of the amplitudes is squared, we may expect it to show considerable oscillation. If $\mathrm{Re}\,b_{1x}$ were independent of q, in fact, it would play precisely the role of the parameter c in Eqs. (1.6) and (1.7), that is to say $2\,\mathrm{Re}\,b_{1x}$ would represent the distance apart of the two diffuse slits that effectively furnish the diffraction pattern. We find in practice that $\mathrm{Re}\,b_{1x}$ given by Eq. (5.32a) doesn't vary very strongly with q, and the distribution of scattered intensities does look a great deal like that of two fixed diffuse slits. But to the extent that $\mathrm{Re}\,b_{1x}$ is indeed found to decrease somewhat with increasing q, the oscillations of the scattered intensity will be found to increase slowly in period as q increases. That is already an indication that the oscillations of the differential cross section do not remain precisely periodic at large momentum transfers.

6

Scattering from Linear Structures

When the target is endowed with two or more scales, the image, or cross section distribution, can get rather involved. Before exploring scattering from various two- and three-dimensional structures, we shall study scattering from one, two, and three parallel one-dimensional slits here. These will however be endowed with some structure, allowing us to exhibit some features of scattering from multiscale objects.

6.1 One Slit

Scattering from an infinite slit is given by the Fourier transform

$$G(q) = \int_{-\infty}^{\infty} \mathrm{d}x\, e^{iqx} A(x), \tag{6.1}$$

obtained by integrating across the slit, and where $A(x)$ is a normalized "transparency distribution",

$$\int \mathrm{d}x\, A(x) = 1. \tag{6.2}$$

Below, we consider a few simple examples of such transparency profiles, restricting ourselves to cases for which the integral in Eq. (6.1) can be evaluated analytically. The functions $A(x)$ that we consider are shown in Fig. 6.1, both in linear and logarithmic representations.

Let us start by first considering a constant profile,

$$A_1(x) = \begin{cases} \dfrac{1}{2\beta}, & |x| \le \beta \\ 0 & \text{outside} \end{cases} \tag{6.3}$$

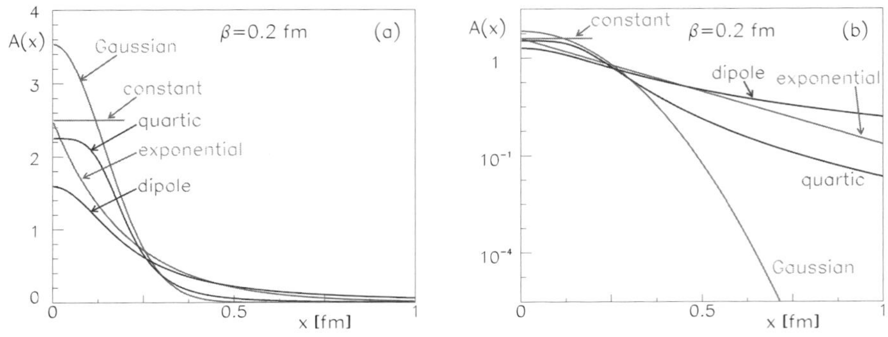

Figure 6.1 Some simple slit profiles, given by Eqs. (6.3), (6.5), (6.7), (6.9), and (6.11). The same functions are shown both in linear (a) and logarithmic (b) representations.

with β characterizing the width. Then

$$G_1(q) = \frac{\sin(q\beta)}{q\beta}. \tag{6.4}$$

The scattering cross section will have a modulation, given by $\sin^2(q\beta)$, and it will be damped at increasing momentum transfer by $1/(q\beta)^2$.

Next, we consider a "dipole" distribution,

$$A_2(x) = \frac{1}{\pi}\frac{\beta}{x^2+\beta^2}, \tag{6.5}$$

for which

$$G_2(q) = \exp(-q\beta). \tag{6.6}$$

This is recognized as the distribution discussed in Chapter 1, see Eq. (1.2). For $q\beta \gg 1$ the cross section will be exponentially damped, as already observed.

Also a quartic distribution,

$$A_3(x) = \frac{\sqrt{2}}{\pi}\frac{\beta^3}{x^4+\beta^4} \tag{6.7}$$

leads to an analytic amplitude

$$G_3(q) = \sqrt{2}\exp(-q\beta\sqrt{2})\sin(q\beta/\sqrt{2}+\pi/4). \tag{6.8}$$

For $q\beta \gg 1$ the cross section will be exponentially damped, but there is a superimposed oscillation, reminiscent of the Fourier transform in Eq. (6.4) of the constant distribution.

Next, we consider a profile that is exponential in shape, but symmetric around its center, $x = 0$:

$$A_4(x) = \frac{1}{2\beta} \exp[-|x|/\beta]. \tag{6.9}$$

One finds

$$G_4(q) = \frac{1}{1 + (q\beta)^2}. \tag{6.10}$$

For $q\beta \gg 1$ the cross section will be damped as a power law.

As a final example, we consider a Gaussian profile,

$$A_5(x) = \frac{1}{\sqrt{\pi}\beta} \exp[-x^2/\beta^2], \tag{6.11}$$

for which the amplitude is also a Gaussian:

$$G_5(q) = \exp[-q^2\beta^2/4]. \tag{6.12}$$

This cross section is dramatically different from the previous ones, it has a Gaussian damping for $q\beta \gtrsim 1$.

We show in Fig. 6.2 the differential cross sections given by these profiles, $G_i^2(q)$, for a common "diffuseness parameter", $\beta = 0.2$ fm. From the fact that the constant and quartic profiles both yield an oscillatory pattern, it is tempting to conclude that this is due to the quartic distribution also having some kind of an edge or shoulder.

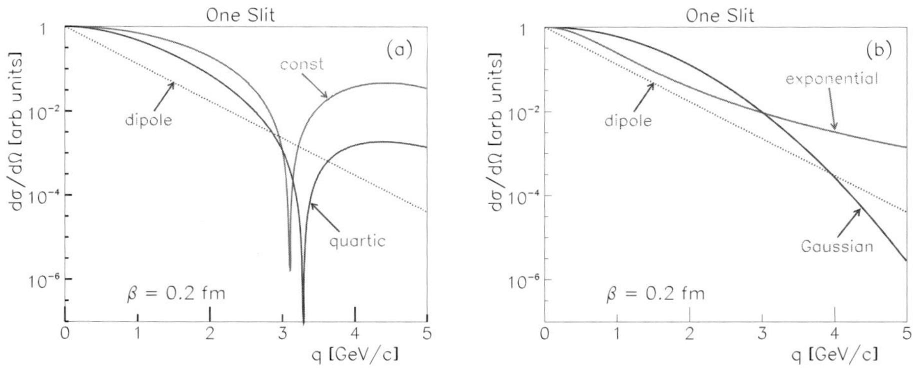

Figure 6.2 Differential cross sections (arbitrary normalization) for scattering from a single slit. (a) solid: constant profile, given by $G_1(q)$ and quartic profile, given by $G_3(q)$; dashed: dipole profile, given by $G_2(q)$. (b) solid: exponential profile given by $G_4(q)$ and Gaussian profile, given by $G_5(q)$; dashed: same (dipole) reference as in (a).

6.2 Two Slits

Scattering from two slits introduces a new scale, namely their separation. Let us consider scattering from two parallel slits, located at $x = \pm R$. Each will contribute a (conveniently normalized) scattering amplitude

$$F_{\pm}(q) = \tfrac{1}{2}\int_{-\infty}^{\infty} \mathrm{d}x e^{iq(\pm R+x)} A(x). \tag{6.13}$$

The total amplitude is given by

$$F(q) = F_{-}(q) + F_{+}(q). \tag{6.14}$$

With two identical slits, the amplitude will factorize,

$$F(q) = \cos(qR)G(q), \tag{6.15}$$

where $G(q)$ is defined by Eq. (6.1) and could be one of the amplitudes given above. This function $G(q)$ represents damping with increasing momentum transfer, and possibly also modulation. We show in Fig. 6.3(a) differential cross sections corresponding to the above profile transparency distributions.

We note that whereas the period of the rapid oscillations is determined by the separation between the two slits, the envelope is determined by the internal transparency distribution of the slits.

This factorization Eq. (6.15) is a result of having identical (and well separated) slits, and leads to zeros in the cross section. If the two slits were different, the amplitude would still have had the same oscillatory pattern, but without zeros. This can be seen as follows. Let

$$\begin{aligned} F(q) &= \tfrac{1}{2}[e^{-iqR}G_i(q) + e^{iqR}G_j(q)] \\ &= \cos(qR)[G_i(q) + G_j(q)] - i\sin(qR)[G_i(q) - G_j(q)]. \end{aligned} \tag{6.16}$$

Then,

$$\begin{aligned} |F(q)|^2 &= \tfrac{1}{4}\{\cos^2(qR)[G_i(q) + G_j(q)]^2 + \sin^2(qR)[G_i(q) - G_j(q)]^2\} \\ &= \tfrac{1}{4}[G_i(q) - G_j(q)]^2 + \cos^2(qR)G_i(q)G_j(q). \end{aligned} \tag{6.17}$$

The regular periodic oscillations are now offset from zero by a q-dependent amount determined by the difference between the two profiles, $\frac{1}{4}[G_i(q) - G_j(q)]^2$. As an example, we show in Fig. 6.3(b) the scattering from the interference of an exponential and a Gaussian slit,

$$F(q) = \tfrac{1}{2}[e^{-iqR}G_4(q) + e^{iqR}G_5(q)]. \tag{6.18}$$

The overall period of oscillation is of course unchanged, but a more complicated pattern is generated. The resulting cross section has minima offset from zero, and values intermediate between the two “pure” distributions.

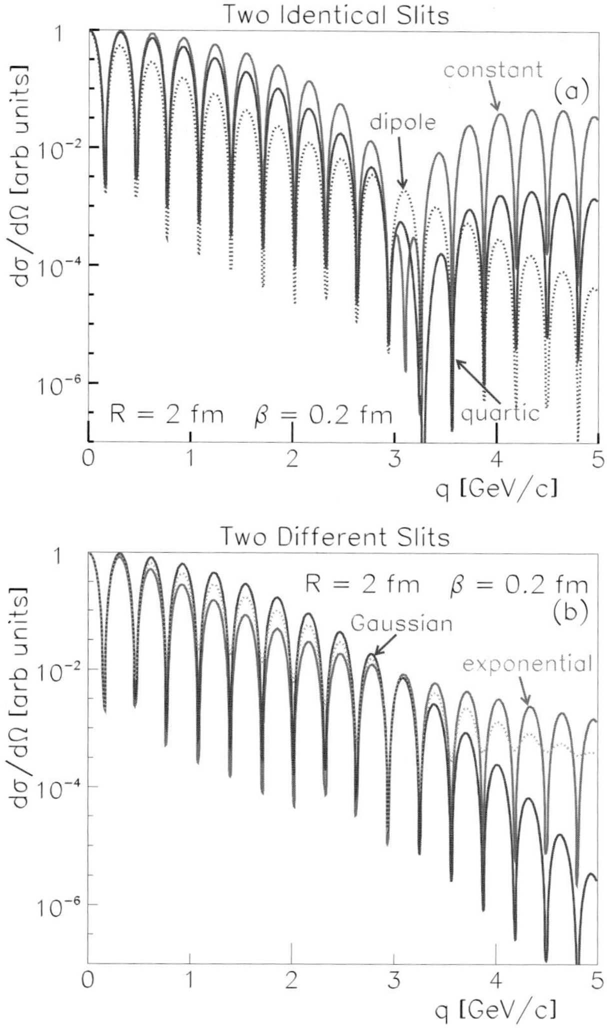

Figure 6.3 Differential cross sections (arbitrary normalization) for scattering from two slits. Different slit transparency profiles are compared. (a) solid: constant profile and quartic profile, Eq. (6.5); dashed: dipole profile, Eq. (6.3). (b) solid: exponential profile, Eq. (6.9); and Gaussian profile, Eq. (6.11); dashed: interference of two different slits, Eq. (6.18).

6.3 Three Slits

Let us first consider three identical and equidistant slits, located at $x = -R$, $x = 0$ and $x = R$. The scattering amplitude (conveniently normalized) can then be expressed as

$$F(q) = \tfrac{1}{3}[F_-(q) + F_0(q) + F_+(q)]. \tag{6.19}$$

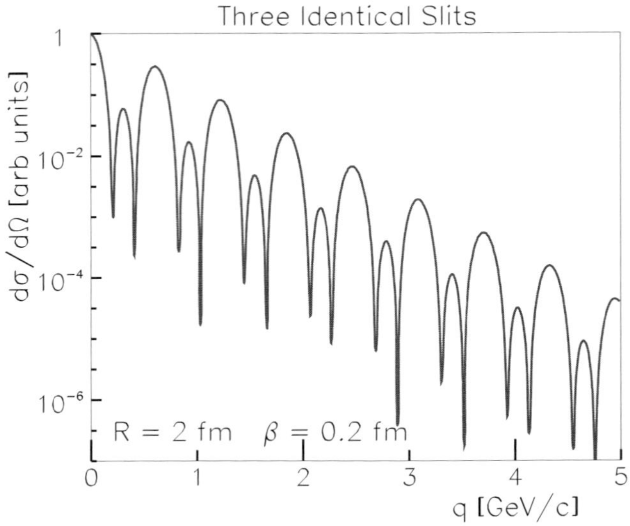

Figure 6.4 Differential cross sections (arbitrary normalization) for scattering from three identical slits. Equidistant slits located at $-2, 0$ and $+2$ fm. The slit profile is given by $G_3(q)$.

It can be taken over from the two-slit results above, with the substitution

$$\cos(qR) \rightarrow \tfrac{1}{3}[1 + 2\cos(qR)], \tag{6.20}$$

leading to

$$\begin{aligned}\frac{|F(q)|^2}{G^2(q)} &= \tfrac{1}{9}[4\cos^2(qR) + 4\cos(qR) + 1] \\ &= \tfrac{1}{9}[2\cos(2qR) + 4\cos(qR) + 3].\end{aligned} \tag{6.21}$$

Note that the inter-slit separation is now R, not $2R$ as in the two-slit case.

The differential cross section is shown in Fig. 6.4, for the case $R = 2$ fm, and the slit profile given by $G_2(q)$. A regular oscillatory pattern is seen, displaying these two periods. Apart from the modulation provided by $G_2^2(q)$, successive maxima alternate in magnitude by a factor of 9 or $1/9$.

In the more general case of three identical slits positioned at x_1, x_2, and x_3, the amplitude can be written as

$$\begin{aligned}F(q) &= \tfrac{1}{3}[e^{iqx_1} + e^{iqx_2} + e^{iqx_3}]G(q) \\ &= \tfrac{1}{3}e^{iq(x_1+x_2+x_3)/3}[e^{iq(2x_1-x_2-x_3)/3} + e^{iq(2x_2-x_1-x_3)/3} \\ &\quad + e^{iq(2x_3-x_1-x_2)/3}]G(q).\end{aligned} \tag{6.22}$$

Apart from an overall (irrelevant) phase, each term contains two factors of the kind $\exp[iz_{ij}/3]$, where

$$z_{ij} = q(x_i - x_j). \tag{6.23}$$

The squared modulus can be written out in terms of trigonometric functions of these arguments, yielding

$$\frac{|F(q)|^2}{G^2(q)} = \tfrac{1}{9}\{2[\cos(z_{12}) + \cos(z_{13}) + \cos(z_{23})] + 3\}. \tag{6.24}$$

In the limit of

$$|z_{12}| = |z_{23}| = qR, \quad |z_{13}| = 2qR, \tag{6.25}$$

we recover the result of Eq. (6.21).

We show in Fig. 6.5 two examples of scattering from nonequidistant slits. In the first example (Fig. 6.5(a)) the three arguments of the trigonometric functions are

$$|z_{12}| = q, \quad |z_{23}| = 3q, \quad \text{and } |z_{13}| = 4q. \tag{6.26}$$

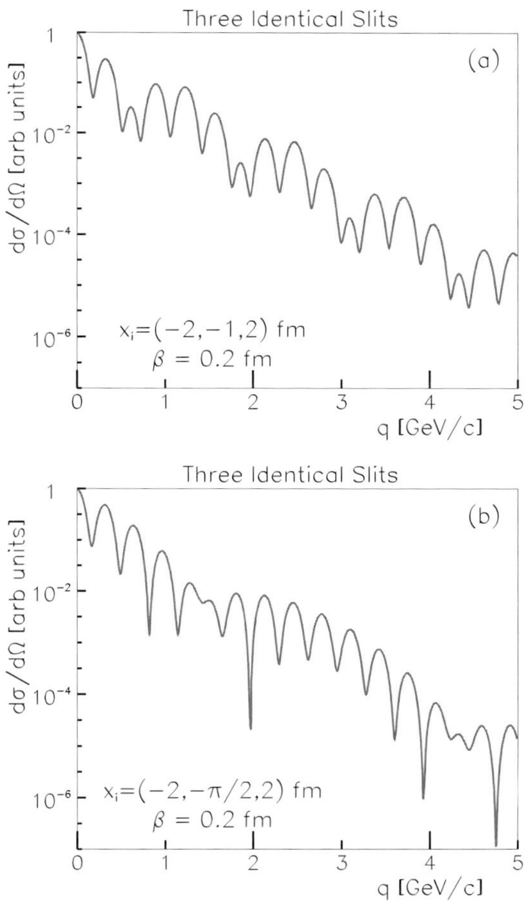

Figure 6.5 Differential cross sections (arbitrary normalization) for scattering from three identical slits. (a) Slits located at $-2, -1$ and $+2$ fm. (b) Slits located at $-2, -\pi/2$ and $+2$ fm. The slit profile is the dipole one, given by $G_2(q)$.

Since the slit separations are related by a rational fraction, $|z_{12}|:|z_{23}| = 1{:}3$ the pattern is repeated.

In the second example (Fig. 6.5(b)) the corresponding arguments of the trigonometric functions are

$$|z_{12}| = \tfrac{1}{2}(4-\pi)q, \quad |z_{23}| = \tfrac{1}{2}(4+\pi)q, \quad \text{and} \quad |z_{13}| = 4q. \tag{6.27}$$

Now, the slit separations are related by an irrational fraction, $|z_{12}|:|z_{23}| = (4-\pi):(4+\pi)$. Thus, there is no repetition of the pattern. However, the short-period oscillation, given by $|z_{13}| = 4q$ is common in these two examples, as is reflected in the illustrations.

The examples shown in this chapter illustrate how multiple scales can lead to more involved cross sections. However, in some cases, the pattern may be sufficiently systematic that it can be decoded in terms of an underlying geometry.

Part II

Phase Shift Functions for Various Radial Falloffs

7

Phase Shift Functions with Gaussian Decrease

The Gausssian function has a number of appealing properties. It is simple and analytic. It is finite in range, but quite smooth in its behavior. Since it happens also to play a fundamental role in constructing the harmonic oscillator wave functions of the nuclear shell model, it is frequently used to represent the densities of the lightest nuclei. It is sometimes used also to represent the short-range interactions between incident particles and such nuclei. If such an interaction potential, $V(\boldsymbol{r})$, is Gaussian in shape, then the phase shift function $\chi(b)$ given by Eq. (2.5) is a Gaussian function of b. We can write it as

$$\chi_{\mathrm{g}}(b_x, b_y) = A_{\mathrm{g}} \exp(-b^2/\beta^2), \tag{7.1}$$

where $b^2 = b_x^2 + b_y^2$, and the strength parameter A_{g} may, in general, be complex. We shall first study the scattering for this phase shift function in some detail, and then discuss two closely related cases in which the phase shift function varies rather less inside a radius $b = c$, but has a similar Gaussian falloff at larger values of b. These are

$$\chi_l(b_x, b_y) = A_\ell \log\{1 + \exp[(c^2 - b^2)/\beta^2]\}, \tag{7.2}$$

$$\chi_{\mathrm{Fg}}(b_x, b_y) = A_{\mathrm{Fg}}/\{1 + \exp[(c^2 - b^2)/\beta^2]\}. \tag{7.3}$$

They are suggestive, to a degree, of phase shift functions that have been used to represent scattering by nuclei with diffuse surfaces.

For illustrative purposes, we choose the relative normalization

$$A_{\mathrm{Fg}} = A_\ell = A_{\mathrm{g}}/\exp[c^2/\beta^2], \tag{7.4}$$

and compare the three phase shift functions in Fig. 7.1, for $A_{\mathrm{g}} = 3, c = \beta = 1$ fm. At first sight (Fig. 7.1(a), where they are plotted linearly) they appear rather different. However, in a typical case of strong absorption, the central region will not contribute very much to the scattering amplitude; instead it is the edge region which

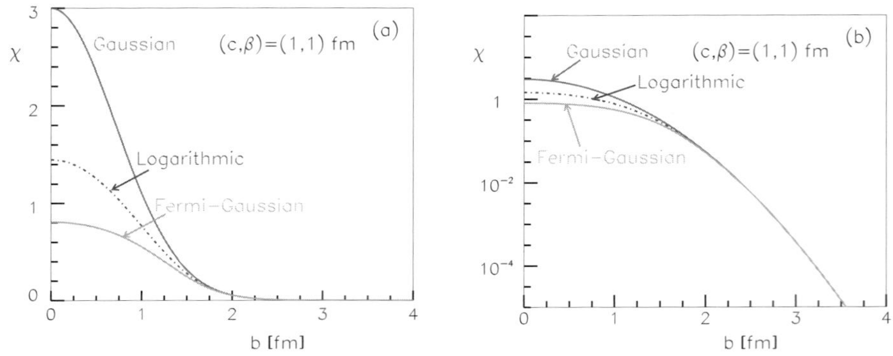

Figure 7.1 Phase shift functions Eqs. (7.1–7.3) with a Gaussian edge. The same three phase shift functions are shown both in linear (a) and logarithmic (b) representations.

is important. This region is better illustrated on a logarithmic scale, in Fig. 7.1(b). With the normalization of Eq. (7.4), the edge region for these three phase shift functions is essentially the same. We will see below, that the differential scattering cross sections will also be rather similar.

Since these phase shift functions are all symmetric under rotation, all the stationary phase points will occur at $b_y = 0$. We will therefore, as in Chapter 5, only need to make use of the function $\chi(b_x, 0)$. It will be convenient therefore to abbreviate it as

$$X(b_x) \equiv \chi(b_x, 0). \tag{7.5}$$

An important difference between the three functions Eqs. (7.1–7.3) is that the Gaussian function $X(b_x)$ is analytic in the entire complex b_x-plane. For the logarithmic case Eq. (7.2) and the "Fermi–Gaussian" Eq. (7.3), on the other hand, the function $X(b_x)$ possesses branch points and poles successively for

$$b_x^2 = c^2 + (2n+1)i\pi\beta^2, \quad n = 0, \pm 1, \pm 2, \ldots \tag{7.6}$$

7.1 The Real-Valued Gaussian Phase Shift Function

For the simple Gaussian phase shift function of Eqs (7.1) and (7.5),

$$X(b_x) = A_g \exp(-b_x^2/\beta^2), \tag{7.7}$$

the stationary points are determined by Eq. (5.2) or

$$X'(b_x) = -2(b_x/\beta^2)A_g \exp(-b_x^2/\beta^2) = q. \tag{7.8}$$

It is instructive, before letting A_g take on complex values, to consider the case of real A_g in some detail. The behavior of the scattering is similar in some ways to that of the simplified model we considered in Section 4.3. It will furnish us with another example of the "rainbow" phenomenon, an angle at which the differential cross section, as we have approximated it, becomes singular. This time, however, we shall go a step further in the approximation procedure and show how the improvement removes the singularity.

To find stationary points we can begin by squaring both sides of Eq. (7.8). If we let

$$Z = \frac{b_x^2}{\beta^2} \tag{7.9}$$

and

$$\mu = \frac{q\beta}{2A_g}, \tag{7.10}$$

then the resulting relation can be written as

$$Ze^{-2Z} = \mu^2. \tag{7.11}$$

For real Z, the expression Ze^{-2Z} has a maximum value $(2e)^{-1}$ at $Z = \frac{1}{2}$. Eq. (7.11) has two real, positive roots, for $0 \leq \mu^2 \leq (2e)^{-1}$. One has $Z < \frac{1}{2}$ and the other $Z > \frac{1}{2}$. Two possible roots for b_x,

$$b_x = \pm\beta\sqrt{Z} \tag{7.12}$$

will correspond to each of the roots for Z, but squaring Eq. (7.8) has in fact given us twice as many roots as we need. Half of them are spurious. It is easy to see that for $A_g > 0$, i.e. an attractive potential, only the negative roots for b_x will actually satisfy Eq. (7.8). Likewise, for $A_g < 0$, only the positive roots will do.

Let us assume, for definiteness, that $A_g > 0$, and that the parameter μ lies in the interval $0 \leq \mu \leq (2e)^{-\frac{1}{2}}$, so that we have

$$0 \leq q \leq \frac{A_g}{\beta}\sqrt{\frac{2}{e}}. \tag{7.13}$$

If we write the two roots of Eq. (7.11) as $Z_1 \geq \frac{1}{2}$ and $Z_2 \leq \frac{1}{2}$, the stationary points we seek are given by (the index β is being reserved for the stationary point $b_{\beta x}$, to be defined below)

$$b_{\alpha x} = -\beta\sqrt{Z_1} \leq -\frac{\beta}{\sqrt{2}} \tag{7.14a}$$

$$b_{\gamma x} = -\beta\sqrt{Z_2} \geq -\frac{\beta}{\sqrt{2}}. \tag{7.14b}$$

As q is increased to the upper bound in the inequality Eq. (7.13) the two stationary points $b_{\alpha x}$ and $b_{\gamma x}$ coalesce at the value $-\beta/\sqrt{2}$, which is analogous to the value $b_{x,\mathrm{R}}$ of Section 4.3. We therefore expect a rainbow singularity to appear in the approximated differential cross section at the q-value

$$q_{\mathrm{R}} = \frac{A_{\mathrm{g}}}{\beta}\sqrt{\frac{2}{e}}. \tag{7.15}$$

At the stationary points we have

$$X(b_{jx}) = A_{\mathrm{g}} e^{-Z_j} = -\frac{q\beta^2}{2b_{jx}} \tag{7.16}$$

$$X''(b_{jx}) = -\frac{2A_{\mathrm{g}}}{\beta^2}(1-2Z_j)e^{-Z_j} = \frac{q}{b_{jx}}\left(1 - 2\frac{b_{jx}^2}{\beta^2}\right) \tag{7.17}$$

The integration Eq. (3.4) may then be carried out along the real b_x-axis. We find that the two stationary points, as in Eq. (3.23), yield the two contributions

$$f_j(\mathbf{k}', \mathbf{k}) = \frac{k}{i}\left\{\frac{-b_{jx}}{qX''(b_{jx})}\right\}^{\frac{1}{2}} e^{i[-qb_{jx}+X(b_{jx})]} \quad (j = \alpha, \gamma) \tag{7.18a}$$

$$= \frac{k\beta}{iq}\left\{\frac{-b_{jx}^2}{\beta^2 - 2b_{jx}^2}\right\}^{\frac{1}{2}} e^{-iqb_{jx}(1+\beta^2/2b_{jx}^2)}. \tag{7.18b}$$

For $q > q_{\mathrm{R}}$, both of the stationary points move off the real axis, but only one of them moves into the lower half-plane, to furnish a convergent integral over the coordinate b_y. To find that stationary point we note that it still obeys Eq. (7.11). Our problem then is to find where in the complex Z-plane the expression Ze^{-2Z} can take on real, positive values. If we let

$$Z = re^{i\theta} \tag{7.19}$$

then Eq. (7.11) becomes

$$re^{i(\theta - 2r\sin\theta) - 2r\cos\theta} = \mu^2, \tag{7.20}$$

and since the expression Ze^{-2Z} must be real-valued we see that the root must lie on the curve

$$2r\sin\theta = \theta \tag{7.21}$$

or

$$r = \frac{\theta}{2\sin\theta}. \tag{7.22}$$

The point $\theta = 0$, $r = \frac{1}{2}$, for example, corresponds to the rainbow point $Z = \frac{1}{2}$. When we substitute this back in Eq. (7.20) we find the equation

$$\mu^2 = \frac{\theta}{2\sin\theta} e^{-\theta\cot\theta}. \tag{7.23}$$

The expression on the right is a monotonically increasing function of θ for $0 \leq \theta \leq \pi$ that rises to ∞ from the rainbow value of $1/2e$. If θ is a positive root of this equation we can write Z, using Eqs. (7.19) and (7.22), as

$$Z_+ = \tfrac{1}{2}(\theta\cot\theta + i\theta), \quad \theta > 0. \tag{7.24}$$

The stationary point that we seek then for $A_g > 0$ is given by

$$b_{1x} = -\beta\sqrt{Z_+}. \tag{7.25}$$

We have had to choose the negative branch of the square root, as before, since the positive root is a false one. If the interaction had the opposite sign, on the other hand, it would have been appropriate to use the positive root, and in that case we would need the negative root of Eq. (7.23) for θ.

There are other stationary points in the lower half-plane, but as in the example considered in Chapter 5, we have no need to use them at this stage or even find them. We can carry out the integration, as in our prior examples, by translating the path into the lower half-plane and passing it through the single stationary point we have just found. Then the values of $X(b_x)$ and $X''(b_x)$ at the stationary point are again given by Eqs. (7.16) and (7.17). It is worth noting that for $q > q_R$, the function X'', evaluated as in Eq. (7.17), has a positive imaginary part. The integrand, as indicated, for example, in Eqs. (3.18a) and (3.21b), has a Gaussian decrease to either side of the stationary point. By passing the contour through the stationary point along a horizontal path, in other words, we are integrating through a maximum of the modulus, and, in a sense, using a compromise between the methods of stationary phase and steepest descent. To the extent that the contribution of a stationary point is approximated by a convergent straight-line integral of the sort we have used, all of these compromises corresponding to different directions of the path furnish the same result. It is not too material then which of the names we attach to the method. But, however we choose to approximate the stationary point contributions, it is worth pointing out one conceptual advantage of the method of steepest descents. It furnishes us an excellent determination of a path for crossing the complex plane by traversing whatever stationary points are necessary for an asymptotic approximation. We can use the path of steepest descent as a guide to the significant stationary points, and then approximate their contribution much as we have been doing.

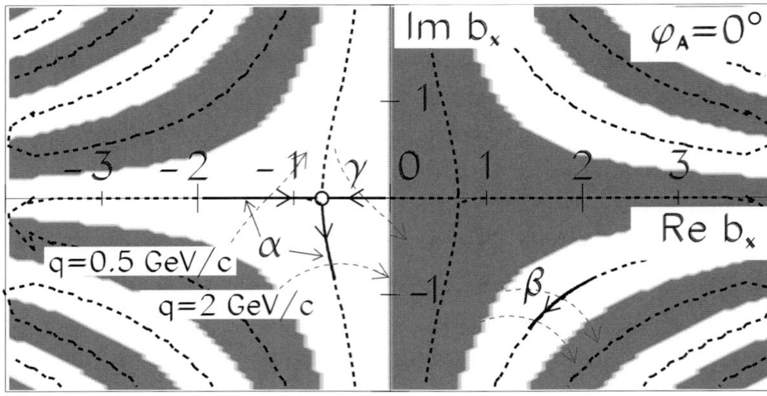

Figure 7.2 Trajectories of stationary points for $X_g(b) = A_g \exp(-b^2/\beta^2)$, with $A_g = |A_g| \exp(i\varphi_A)$, for the real case, $\varphi_A = 0°$. The scale parameter β is chosen as the unit of length. These trajectories are shown in the complex b_x-plane, where $\boldsymbol{b} = (b_x, b_y)$. The thinly dashed curves indicate paths of integration. (At the stationary points, $b_y = 0$.)

In order to develop some better understanding of how the integrand varies in the complex b_x-plane, and of the approximations involved, we display in Fig. 7.2 some properties of $X'(b_x)$. Even without the explicit formulas for the stationary points given above, one can easily determine graphically where they are constrained to lie. Our sign convention is such that stationary-phase points can only occur where $\mathrm{Re}\, X'(b_x) > 0$. As a first step, the complex b_x-plane is therefore divided into regions where $\mathrm{Re}\, X'(b_x) > 0$ (white) alternating with regions where $\mathrm{Re}\, X'(b_x) < 0$ (shaded). Next, we note that at the stationary points, $X'(b_x)$ must be purely real. Thus, we also show the locations where $\mathrm{Im}\, X'(b_x)$ vanishes. These are the dashed lines running along the white and shaded "bands", roughly in their middle, including in this case the negative, real axis. From the above discussion, it is now clear that stationary points can only be found along those dashed lines that run somewhere near the middle of the *white* bands. It may be noted that these are independent of the strength of the interaction, but they do of course depend on the phase; in this example A_g is real.

Indeed, the "trajectories", given by Eq. (5.2) above, which are the paths in the complex b_x-plane, followed by the stationary points as functions of the momentum transfer, q, are shown as solid segments along such dashed lines. Here, the strength of the interaction plays a role, as does the momentum transfer. On the negative real axis, we identify the points α and γ. The segments shown correspond to the range in momentum transfer 0.1 GeV/c $\leq q \leq$ 3 GeV/c. The rainbow point, where for real A_g the trajectories α and γ run together, is shown as an open circle.

Paths of integration corresponding to $A_g = 5$ are shown schematically for $q = 0.5$ and 2 GeV/c (thinly dashed). For the purpose of this illustration, these are determined by mapping out the path of steepest descent on each side of the stationary point. It now becomes clear, that in order to continue the path of integration out to $\mathrm{Re}\, b_x = \infty$, more "ridges" have to be passed. The dominant one of these, β, is also shown. The contribution from this stationary point will be omitted in our discussion of the cross section for the real Gaussian, but it will become more prominent when in Section 7.3 we consider complex interactions, $\mathrm{Im}\, A_g > 0$.

The single stationary point on the integration path for $q > q_R$ yields a scattering amplitude of the same general form Eq. (7.18b) as those for $q < q_R$. Like both of those, it also possesses a singularity at the rainbow point $q = q_R$. For $q < q_R$ the amplitudes furnished by the two stationary points interfere. For $q > q_R$, on the other hand, there is only one such amplitude and it decreases rapidly in modulus as q increases further.

To evaluate our scattering amplitude numerically we should note that their dependence on the four parameters k, q, A_g, and β possesses a certain scale invariance. Because $f(\boldsymbol{k}',\boldsymbol{k})$ is simply proportional to k, and A_g is dimensionless we can write

$$f(\boldsymbol{k}',\boldsymbol{k}) = k\beta^2\Phi(q\beta, A_g), \tag{7.26}$$

where Φ is the same function for all phase shifts that can be expressed in the Gaussian form, Eq. (7.1). If we write the differential cross section, furthermore in the form

$$\frac{\mathrm{d}\sigma}{\mathrm{d}q^2} = \frac{\pi}{k^2}\frac{\mathrm{d}\sigma}{\mathrm{d}\Omega} = \frac{\pi}{k^2}|f(\boldsymbol{k}',\boldsymbol{k})|^2, \tag{7.27}$$

that has become customary in high-energy elementary particle physics, then we have

$$\frac{\mathrm{d}\sigma}{\mathrm{d}q^2} = \pi\beta^4|\Phi(q\beta, A_g)|^2. \tag{7.28}$$

In that case $\beta^{-4}\,\mathrm{d}\sigma/\,\mathrm{d}q^2$ becomes simply a function of two parameters.

We have graphed the differential cross section $\mathrm{d}\sigma/\,\mathrm{d}q^2$ for $A_g = 5$, 10, and 15, and for $\beta = 1$ fm, in Figs. 7.3(a)–(c). We have adopted the value of the momentum as $k = 1.7$ GeV/c, corresponding to $T_{\mathrm{kin}} = 1$ GeV, and recall that $[\beta = 1\ \mathrm{fm}]^{-1}$ corresponds to a momentum transfer of 0.197 GeV/c.

The dashed curves on the same plots are the differential cross sections for the same parameters that result from a more direct numerical integration of the diffraction amplitude, which, according to Eq. (2.2), is given by

$$f(\boldsymbol{k}',\boldsymbol{k}) = ik\int_0^\infty J_0(qb)\{1 - e^{i\chi(b)}\}b\,\mathrm{d}b, \quad q = |\boldsymbol{k}' - \boldsymbol{k}|. \tag{7.29}$$

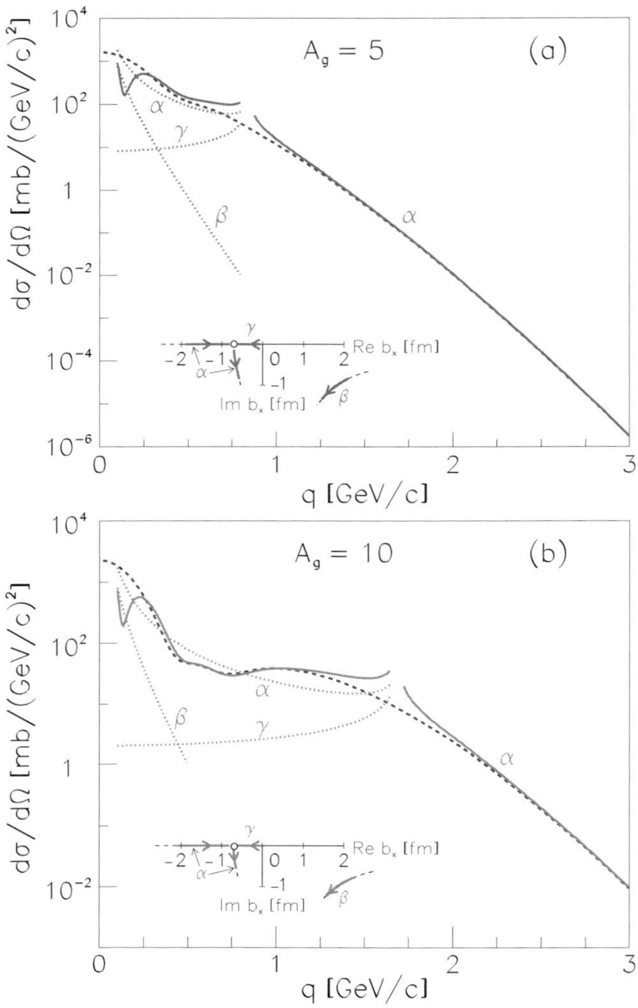

Figure 7.3 Differential cross sections for $\chi_g(b) = A_g \exp(-b^2/\beta^2)$, with A_g real; (a) $A_g = 5$, (b) $A_g = 10$, (c) $A_g = 15$ (next page). The scale parameter β is 1 fm. The inserts show the trajectories of stationary points for the range in momentum transfer $0.02\ \text{GeV/c} \leq q \leq 6\ \text{GeV/c}$ (dotted). The parts of the trajectories that correspond to the cross section plots, $0.1\ \text{GeV/c} \leq q \leq 3\ \text{GeV/c}$, are indicated as solid curves. The arrows indicate the sense in which the stationary points move as the momentum transfer increases. The differential cross sections contributed by each of the stationary points, α, β, and γ, are shown as dotted curves. The trajectory γ contributes only out to the rainbow point. The solid curves show the resulting cross section, evaluated in the asymptotic approach (rainbow singularities occur at the discontinuities), whereas the dashed curves show the results of numerical evaluations of the diffraction integrals.

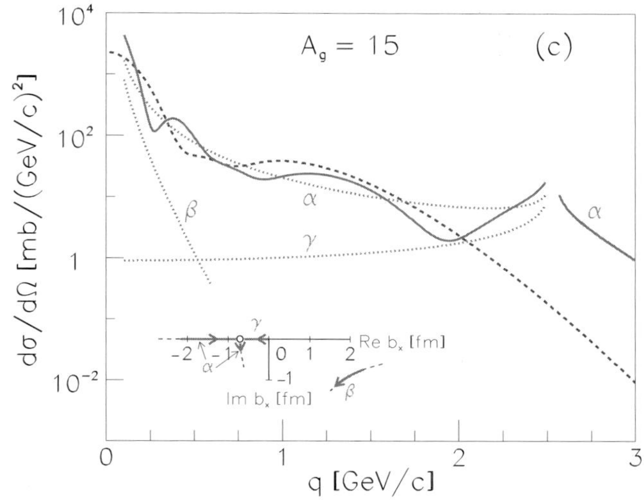

Figure 7.3 (c) See caption to parts (a) and (b) on previous page.

For a phase shift function that has a restricted range, as is the case with the Gaussian, this integral is easily evaluated, and the agreement is seen to be excellent, except for q-values in two limited ranges. One of these is the range of small q-values, $q\beta \ll 1$, to which the asymptotic approximation simply does not apply. The other is the region near the rainbow point q_{R}, where the approximation becomes singular, and the true cross section is not (for the three values of A_g, the rainbow points are located at 0.85, 1.70, and 2.54 GeV/c). We shall discuss the behavior near the rainbow point a bit further shortly, but let us first note some of the broader features of these results.

The wavy appearance of the differential cross section for $q < q_{\mathrm{R}}$, i.e., inside the rainbow, clearly results from the interference of the amplitudes provided by the two stationary points on the real b_x-axis. For $q > q_{\mathrm{R}}$, on the other hand the smooth falloff is due to the amplitude contributed by the single stationary point. As we have noted earlier, it is the exponential function in the amplitude Eq. (7.18b) that dominates its q-dependence; the other factors have a much weaker dependence on q. The squared modulus of this exponential function alone yields

$$|f_j(\boldsymbol{k}',\boldsymbol{k})|^2 \sim \exp\left\{2q\,\mathrm{Im}\left[b_{jx}\left(1+\frac{\beta^2}{2b_{jx}^2}\right)\right]\right\}. \tag{7.30}$$

Since the stationary point for $q > q_{\mathrm{R}}$ has $\mathrm{Im}\, b_x$ negative, and since $|b_{jx}|^2 > \frac{1}{2}\beta^2$ we can recognize a rate of decrease that is essentially exponential in q. Of course the stationary point b_x itself will also depend somewhat on the value of q, so we can say a bit more about the q-dependence of the expression Eq. (7.30) by seeing how $\operatorname{Im} b_{jx}$ varies for large q. For quite large momentum transfers, $q \gg q_R$, the root of Eq. (7.23) is

$$\theta \sim \pi\left(1 - \frac{1}{2\log\mu}\right), \tag{7.31}$$

which shows that

$$Z_+ \sim -\log\mu + \frac{i\pi}{2}, \tag{7.32}$$

and

$$\begin{aligned}\operatorname{Im} b_x &\sim -\beta\sqrt{\log\mu} \\ &\sim -\beta\sqrt{\log\left(\frac{q\beta}{2A_g}\right)}.\end{aligned} \tag{7.33}$$

The differential cross section should therefore show a very slow increase in its logarithmic slope with increasing q, and that is indeed evident in the graphs.

7.2 Differential Cross Section near the Rainbow Point

Our approximation to the differential cross section, as we have seen, diverges at the rainbow point $q = q_R$. That is because our method of approximation is an asymptotic one, valid in the limit of short wavelengths. In the limit of zero wavelength there is indeed a sharp singularity in the physical cross section, but the approach to that limit is slow. For the cases we have graphed $\beta q_R = A_g\sqrt{2/e}$ ranges from only 4.3 to 12.9, and the numerically integrated diffraction amplitude shows little hint of that behavior. For such modest values of βq_R, therefore, there may be some value in correcting our approximation procedure, at least in the neighborhood of the rainbow point.

Our problem is to find a more sensitive way of carrying out the integration Eq. (3.4) when the stationary points on which we have thus far based it are quite near the point $b_{x,R}$ at which

$$X''(b_{x,R}) = \frac{\partial^2}{\partial b_x^2}\chi(b_x,0)\bigg|_{b_x=b_{x,R}} = 0. \tag{7.34}$$

Since the function $\chi(b)$ is rotationally invariant, its value for any b_x is stationary as a function of b_y for $b_y = 0$. We can always expand $\chi(b_x, b_y)$ about $b_y = 0$ and carry out the b_y integration as we did in Eq. (3.21a). It is the integration over b_x that we shall have to improve. We can do that by expanding $X(b_x) = \chi(b_x, 0)$ not about the stationary points we have used before, but instead about the point $b_{x,R}$ at which X'' vanishes. If we carry out this expansion to third order we have

$$X(b_x) = X(b_{x,R}) + (b_x - b_{x,R})X'(b_{x,R}) + \tfrac{1}{6}(b_x - b_{x,R})^3 X'''(b_{x,R}). \tag{7.35}$$

The integral that we must evaluate is then of the form

$$J(q) = \lim_{\epsilon \to 0} \int_{-\infty}^{\infty} \exp\{-\epsilon b_x^2 + i\xi(b_x - b_{x,\mathrm{R}}) + i\eta(b_x - b_x R)^3\}\, \mathrm{d}b_x, \tag{7.36}$$

which is convergent for $\epsilon \to 0$. If we shift the coordinate b_x we can then write

$$J(q) = 2\int_0^{\infty} \cos\{\eta b_x^3 + \xi b_x\}\, \mathrm{d}b_x. \tag{7.37}$$

The coeffficients η and ξ in this expression are given by

$$\xi = q - X'(b_{x,\mathrm{R}}) = q - q_{\mathrm{R}} \tag{7.38}$$

$$\eta = -\tfrac{1}{6} X'''(b_{x,\mathrm{R}}). \tag{7.39}$$

We have here made use of the fact that the integrand in Eq. (7.37) is unchanged under a simultaneous change of sign of η and ξ, and reversed the signs of both η and ξ in the identification above.

The integral Eq. (7.37) is an Airy function [1],

$$J(q) = \frac{2\pi}{(3\eta)^{\frac{1}{3}}} \mathrm{Ai}\left(\frac{\xi}{(3\eta)^{\frac{1}{3}}}\right). \tag{7.40}$$

Its behavior is not at all singular at the rainbow point. The full scattering amplitude in this approximation is given by

$$f(\boldsymbol{k}', \boldsymbol{k}) = \frac{k}{i}\left\{\frac{2\pi i b_{x,\mathrm{R}}}{q}\right\}^{\frac{1}{2}} \left(\frac{2}{-X'''(b_{x,\mathrm{R}})}\right)^{\frac{1}{3}} \mathrm{Ai}\left[(q - q_{\mathrm{R}})\left(\frac{2}{-X'''(b_{x,\mathrm{R}})}\right)^{\frac{1}{3}}\right]$$
$$\times\, e^{-iqb_{x,\mathrm{R}} + iX(b_{x,\mathrm{R}})}. \tag{7.41}$$

The Cartesian derivative of $X(b_x)$ in this expression can be written in terms of radial derivatives of $\chi(b)$ for real values of b_x as

$$X'''(b_x) = \frac{b_x}{|b_x|} \chi'''(|b_x|). \tag{7.42}$$

Since the expression Eq. (7.41) for the scattering amplitude is based on an expansion about the point $b_{x,\mathrm{R}}$, which is $\beta/\sqrt{2}$ for the Gaussian phase shift, it can not be expected to hold over a broad range of q-values. As we can see from the differential cross section curves shown in Fig. 7.4, the expression based on our third order expansion does remove the singularity at the rainbow point, and substantially improves the asymptotic approximation in the region about it.[1]

[1] The Airy result represents the contributions from the stationary points α and γ. The contribution of the β trajectory would give some additional contribution inside the rainbow.

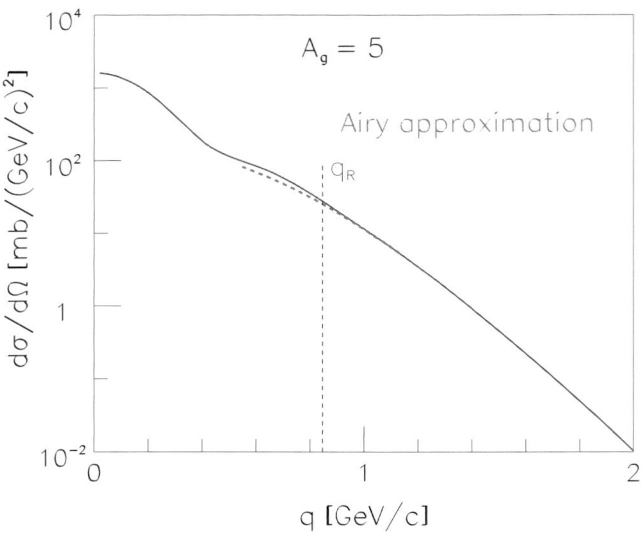

Figure 7.4 Differential cross sections for $\chi_g(b) = A_g \exp(-b^2/\beta^2)$, with $A_g = 5$ and $\beta = 1$ fm. The solid curve shows the results of a numerical integration of the diffraction integral, while the dashed curve shows the result of the Airy approximation.

While we have taken some pains to correct the asymptotic approximation in the neighborhood of the rainbow points, we should note that such corrections are not often necessary. Rainbow points are rarities in the complex plane. They tend to occur only for real-valued phase shift functions, and then only on the real axis. When $X(b_x)$ is a complex-valued analytic function, its second derivative, i.e., both real and imaginary parts, will not in general vanish at any of the stationary points we have defined earlier. It follows then, that rainbow scattering, as we have defined it, tends not to occur in the purely nuclear domain. However, as we shall see in Chapters 11 and 12, for scattering from an extended charge, without or even with additional nuclear interactions, there may be rainbow phenomena implicit.

7.3 Gaussian Phase Shift Function with Absorption

When absorption is present, we must let A_g become complex,

$$A_g = |A_g| e^{i\varphi_A}, \quad 0 \le \varphi_A \le \pi . \tag{7.43}$$

Also in this more general case, the stationary points are determined by Eq. (7.8), but the solutions b_{jx} will now be complex, even for small values of the momentum transfer. By taking the logarithms on both sides of Eq. (7.8), the equation for the stationary points can then be written as

$$b_{jx}^2 = \beta^2\left[\log\left(\frac{2A_g}{q\beta}\right) + \log\left(\frac{-b_{jx}}{\beta}\right)\right]. \tag{7.44}$$

If we further write b_{jx} as

$$b_{jx} = |b_{jx}|e^{i\varphi}, \qquad -\pi \le \varphi \le 0, \tag{7.45}$$

then Eq. (7.44) is equivalent to

$$\cos(2\varphi) + \frac{\sin(2\varphi)}{2[\varphi_A + \varphi + (2n+1)\pi]}\left\{\log\left(\frac{\sin(2\varphi)}{2[\varphi_A + \varphi + (2n+1)\pi]}\right) - \log\left[2\left(\frac{|A_g|}{q\beta}\right)^2\right]\right\} = 0, \quad \text{with } n = 0, \pm 1, \pm 2, \ldots, \tag{7.46}$$

and with

$$b_{jx} = \beta\sqrt{\left|\frac{\varphi_A + \varphi + (2n+1)\pi}{\sin(2\varphi)}\right|}\, e^{i\varphi}. \tag{7.47}$$

We note that Eq. (7.44) for the complex variable b_{jx} has effectively been reduced to an equation for *one* real variable φ, since Eq. (7.46) is independent of the modulus $|b_{jx}|$. The relationship, Eq. (7.47), between the phase φ and the modulus $|b_{jx}|$ is independent of $|A_g|$ and q, in agreement with the general statements, Eqs. (3.5) and (7.8).

Points of stationary phase are shown in Fig. 7.5 for $\varphi_A = 30°$ and $90°$; compare Eq. (7.43). It is instructive to compare this with Fig. 7.2 for the real case. We note that, whereas the remaining points of stationary phase in the third quadrant, $b_{\alpha x}$, have moved downward, those in the fourth quadrant, $b_{\beta x}$, have moved upward by a similar distance. While the stationary point in the fourth quadrant was of little importance for real, positive A_g, since it was far from the real axis, it begins to increase in importance as φ_A increases. When $\varphi_A = 90°$, the stationary points $b_{\alpha x}$ and $b_{\beta x}$ are situated symmetrically with respect to the imaginary axis, $b_{\beta x} = -b_{\alpha x}^*$. This is a consequence of the Schwarz reflection symmetry obeyed by $f_j(\boldsymbol{k}', \boldsymbol{k})$, and leads to

$$f_\beta(\boldsymbol{k}', \boldsymbol{k}) = -f_\alpha^*(\boldsymbol{k}', \boldsymbol{k}), \quad \varphi_A = 90°, \tag{7.48}$$

in agreement with the more general result, Eq. (2.11).

These changes in the behavior of the stationary points play an essential role in explaining the changing structure of the cross section as a function of φ_A. Two such differential cross sections and the positions of the associated stationary points are shown in Fig. 7.6. For $\varphi_A < 90°$, the contributions from $b_{\alpha x}$ and $b_{\beta x}$ are seen to be quite different,

$$|f_\beta(\boldsymbol{k}', \boldsymbol{k})| < |f_\alpha(\boldsymbol{k}', \boldsymbol{k})|, \tag{7.49}$$

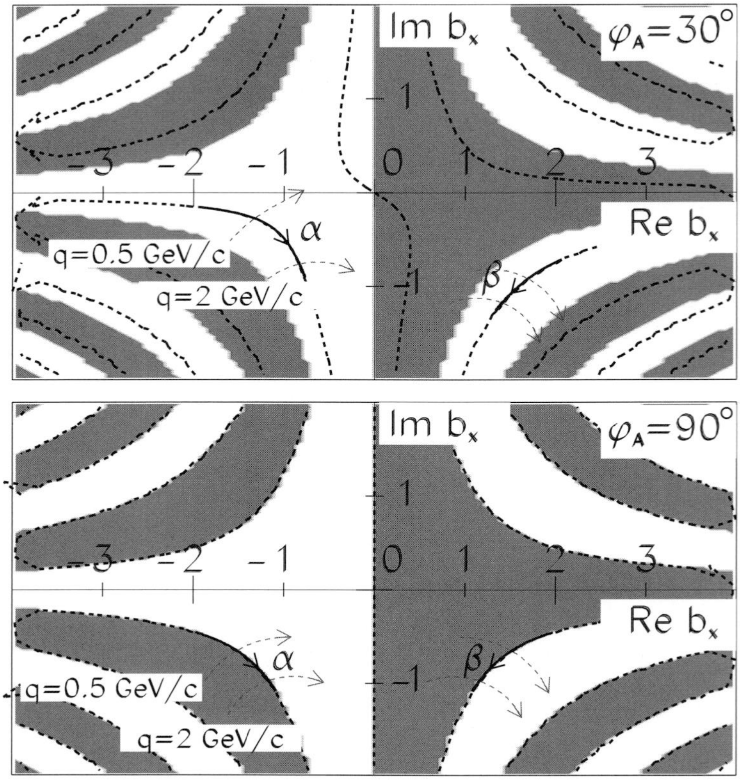

Figure 7.5 Trajectories of stationary points for $X_g(b) = A_g \exp(-b^2/\beta^2)$, with $A_g = |A_g| \exp(i\varphi_A)$, for two values of φ_A: 30° and 90°. Similar to Fig. 7.2.

with $|f_\alpha(\boldsymbol{k}', \boldsymbol{k})|^2$ becoming the dominant contribution to the cross section as $q \to \infty$. On the other hand, for $\varphi_A = 90°$, the two contributions are equal in magnitude [cf. Eq. (7.48)]; the differential cross section has zeros when they interfere destructively.

The variation of the phase of each term with increasing momentum transfer is roughly given by $q \,\mathrm{Re}\, b_{jx}$. This leads to a regular diffractive interference pattern, with period determined by

$$\Delta q = \frac{\pi}{\left|\mathrm{Re}\left\{b_{\alpha x} - b_{\beta x} + \dfrac{1}{q}\left[X(b_{\beta x}) - X(b_{\alpha x})\right]\right\}\right|}$$
$$= \frac{\pi}{\left|\mathrm{Re}\left\{b_{\alpha x}\left[1 + \dfrac{\beta^2}{2|b_{\alpha x}|^2}\right] - b_{\beta x}\left[1 + \dfrac{\beta^2}{2|b_{\beta x}|^2}\right]\right\}\right|}. \tag{7.50}$$

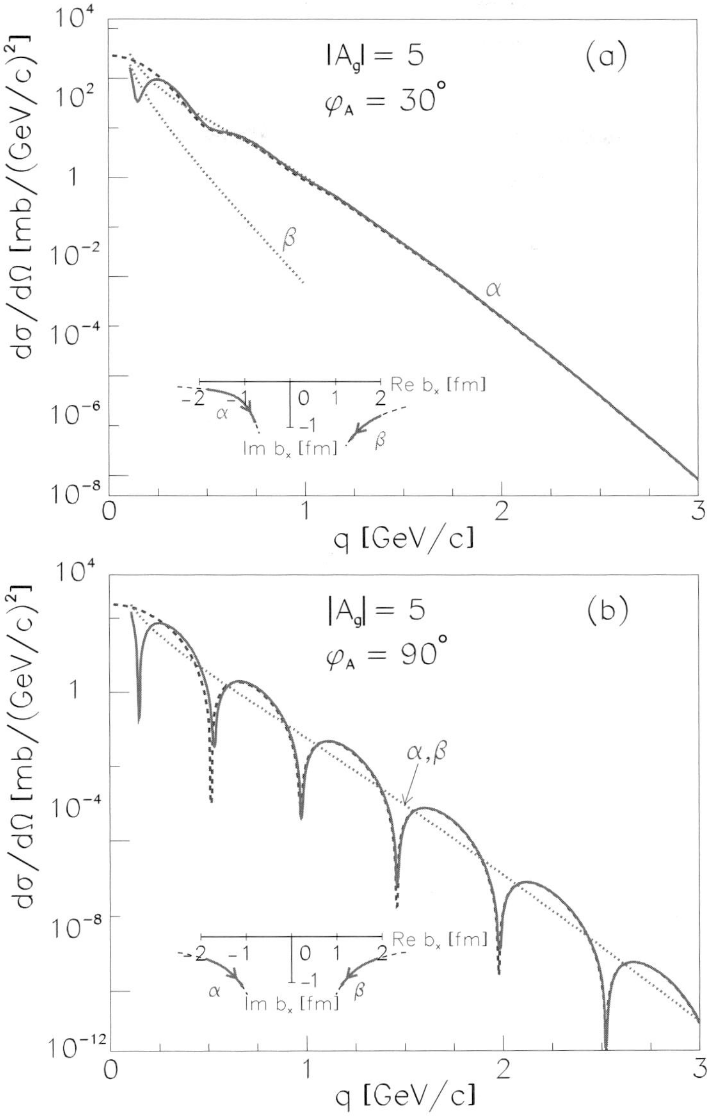

Figure 7.6 Differential cross sections for $\chi_g(b) = A_g \exp(-b^2/\beta^2)$, with $A_g = |A_g| \exp(i\varphi_A)$, $\beta = 1$ fm and $|A_g| = 5$; (a) $\varphi_A = 30°$, (b) $\varphi_A = 90°$. As in Fig. 7.3, the inserts show the trajectories of stationary points, and their individual contributions to the differential cross sections are shown as dotted curves. The results of numerical evaluations of the diffraction integrals are shown as dashed curves.

Asymptotically, the period of oscillation will change very slowly, according to

$$|\operatorname{Re} b_{jx}| \simeq \frac{\pi\beta}{2}\left[\log\left(\frac{q\beta}{2|A_g|}\right) + \mathcal{O}\left(\log\log\left(\frac{q\beta}{2|A_g|}\right)\right)\right]^{-1/2}, \quad (\text{for } \varphi_A = \pi/2). \tag{7.51}$$

Asymptotically, the contribution to the amplitude from each region of stationary phase, b_{jx}, decreases with increasing q roughly as

$$|f_j(\boldsymbol{k}', \boldsymbol{k})| \sim \exp\left\{q \operatorname{Im} b_{jx}\left[1 - \frac{\beta^2}{2|b_{jx}|^2}\right]\right\}, \tag{7.52}$$

since the remaining factors have a much weaker q-dependence. The fact that $\operatorname{Im} b_{\alpha x}$ is negative and decreases further with increasing q, accounts for the steepening slope of the cross section for $q > q_R$. In fact, asymptotically, we find that

$$\operatorname{Im} b_{jx} \simeq -\beta\left[\log\left(\frac{q\beta}{2|A_g|}\right) + \mathcal{O}\left(\log\log\left(\frac{q\beta}{2|A_g|}\right)\right)\right]^{-1/2}, \tag{7.53}$$

and the slope of the cross section will thus increase rather slowly.

In Fig. 7.7 we compare the cross sections obtained for purely absorptive interactions of different strengths. These correspond to the three values of $|A_g| = 1, 5,$

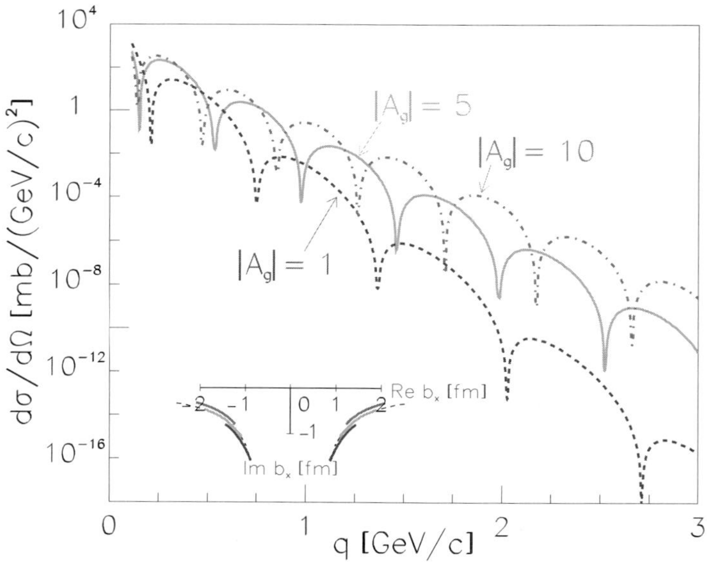

Figure 7.7 Differential cross sections for $\chi_g(b) = i|A_g|\exp(-b^2/\beta^2)$, for $\beta =$ 1 fm. Three values of $|A_g|$ are considered: 1, 5, and 10. The locations of the stationary points (those for $|A_g| = 1$ and 10 are a little offset vertically, for clarity), and the periods and slopes of the envelopes of the differential cross sections are seen to depend on the strengths of the interaction in an essential fashion.

and 10, and $\varphi_A = 90°$ in all three cases. The different periods of oscillation can be understood in terms of Eq. (7.50): the stronger the absorption, the shorter the period of oscillation, in terms of q. Also, for stronger absorption, the slope of the cross section envelope is seen to undergo a decrease, since asymptotically,

$$\operatorname{Im} b_{jx} \simeq -\beta\left[\log\left(\frac{q\beta}{2|A_{\mathrm{g}}|}\right) + \mathcal{O}\left(\log\log\left(\frac{q\beta}{2|A_{\mathrm{g}}|}\right)\right)\right]^{-1/2}. \tag{7.54}$$

The relationship between the classical concept of force and the condition of stationary phase [cf. Eq. (3.5)], is a particularly simple one for the Gaussian phase shift function. The potential corresponding to the Gaussian phase shift function, Eq. (7.1), is readily found from Eq. (2.6c) to be

$$V(\boldsymbol{b}, z) = -\frac{\hbar v}{\sqrt{\pi}\beta}\,\chi(\boldsymbol{b})e^{-z^2/\beta^2}, \tag{7.55}$$

where v is the projectile velocity. At the point of stationary phase, where $\nabla\chi(\boldsymbol{b})\big|_{\boldsymbol{b}=\boldsymbol{b}_j} = \boldsymbol{q}$, the transverse component of the force, $\boldsymbol{F}_t$, is given by

$$\boldsymbol{F}_t(\boldsymbol{b}_j, z) = -\boldsymbol{\nabla}_{\boldsymbol{b}} V(\boldsymbol{b}, z)\big|_{\boldsymbol{b}=\boldsymbol{b}_j} = \frac{v}{\sqrt{\pi}\beta}\boldsymbol{q}\,e^{-z^2/\beta^2}, \tag{7.56}$$

and the value of the potential according to Eq. (7.16) may be written as

$$V(\boldsymbol{b}_j, z) = \frac{v\beta}{2\sqrt{\pi}}\,\frac{\boldsymbol{q}\cdot\boldsymbol{b}_j^*}{|\boldsymbol{b}_j|^2}\,e^{-z^2/\beta^2}. \tag{7.57}$$

Three remarks are here in order:

i. The deflective force, $\boldsymbol{F}_t$, is real along the entire path, $-\infty < z < \infty$, of stationary phase. This is a special property of the Gaussian potential, for which the dependence on transverse and longitudinal coordinates factorizes.
ii. A real potential, Eq. (7.55), will acquire an absorptive part at the stationary impact parameter if that point lies off the real axis, $\operatorname{Im} b_{jx} < 0$. That explains why the cross section tends to fall exponentially beyond the rainbow point, $q > q_{\mathrm{R}}$.
iii. The sign of the real part of the potential for the path corresponding to the stationary impact vector $\boldsymbol{b}_j$ is determined by $\operatorname{Re}(\boldsymbol{q}\cdot\boldsymbol{b}_j)$. Thus, the sign of $\operatorname{Re} V(\boldsymbol{b}_j, z)$ may differ from its value taken for real arguments, given by Eq. (7.55). Two complex paths can thus pass through opposite sides of the nucleus (i.e., with opposite signs of $\operatorname{Re} b_x$), but still lead to the same point of observation. The effective potential, given by Eq. (7.57), can thus be attractive on one side of the nucleus and repulsive on the other. That is in fact why there can be two such opposite trajectories that contribute interfering amplitudes for a given momentum transfer.

Much of the remainder of this Chapter 7 deals with phase shift functions with Gaussian edges, and the somewhat flatter central dependence of Eqs. (7.2) and (7.3). It may be considered as background material that can be skipped in a first reading, in order to focus on other significant features of the phase shift function.

7.4 Logarithmic Phase Shift Function

We consider next the logarithmic phase shift function of Eq. (7.2),

$$\chi_l(b_x, b_y) = A_\ell \log\{1 + \exp[(c^2 - b^2)/\beta^2]\}. \tag{7.58}$$

The equation for the stationary points, $\nabla\chi(\boldsymbol{b}) = \boldsymbol{q}$ can be reduced to

$$\begin{aligned} b_{jx}^2 &= c^2 + \beta^2 \log\left[-1 - \frac{2b_{jx}A_\ell}{q\beta^2}\right] \\ &= c^2 + \beta^2\left\{(2n+1)i\pi + \log\left[1 + \frac{2b_{jx}A_\ell}{q\beta^2}\right]\right\}, \quad \text{for } n = 0, \pm 1, \pm 2, \ldots \end{aligned} \tag{7.59}$$

Asymptotically, as $q \to \infty$ the relevant solutions will approach the logarithmic singularities of $\chi_l(b_x)$ that are located at

$$b_x = \pm\sqrt{c^2 \mp i\pi\beta^2}. \tag{7.60}$$

Eq. (7.59) can easily be solved by using the iteration method discussed in Appendix B, starting at these singularities.

At the stationary points, we have

$$\begin{aligned} X_l(b_{jx}) &= -A_\ell \log\left(1 + \frac{q\beta^2}{2b_{jx}A_\ell}\right), \\ X_l''(b_{jx}) &= -\frac{2qb_{jx}}{\beta^2}\left(1 - \frac{\beta^2}{2b_{jx}^2} + \frac{q\beta^2}{2b_{jx}A_\ell}\right). \end{aligned} \tag{7.61}$$

For an appropriate choice of parameters, the edge of the phase shift function Eq. (7.58) will be the same as that of the Gaussian, i.e., we have

$$X_l(b_x \gg c) \simeq X_g(b_x), \tag{7.62}$$

provided

$$A_\ell\, e^{c^2/\beta^2} = A_g, \quad \beta_l = \beta_g = \beta. \tag{7.63}$$

At small momentum transfers, only the edge of the phase shift function is relevant for the scattering amplitudes. This may be seen quantitatively as follows: if $q\beta/A_\ell$ is sufficiently small, Eq. (7.59) for the stationary points reduces to

Eq. (7.44) which applies for the Gaussian case. The points of stationary phase will thus be the same for the two phase shift functions. It is also readily seen from Eq. (7.61) that for small values of $q\beta/A_\ell$, $X_l(b_{jx})$ and $X_l''(b_{jx})$ will become the same as $X_g(b_{jx})$ and $X_g''(b_{jx})$, respectively. The amplitudes will thus be the same for sufficiently small values of $q\beta/A_\ell$ in the two cases.

At large momentum transfers, however, the two phase shift functions will lead to differential cross sections that are considerably different. Because of the presence of a (logarithmic) singularity, X_l will lead to a more stable pattern of oscillations than that resulting from X_g. This increased stability is illustrated in Fig. 7.8 where we compare the cross sections which correspond to X_l and X_g, for two positions of the singularity.

As a common reference, we have in Fig. 7.8(a) chosen the differential cross section resulting from X_g with $A_g = 3i$ and $\beta = 1$ fm. The parameters considered in Fig. 7.8(a) are such that the singularity is at

$$b_x = \pm\sqrt{1 \mp i\pi} \text{ fm} \quad (\text{or, equivalently}, c = 1 \text{ fm}),$$

whereas those considered in Fig. 7.8(b) are such that the singularity is at

$$b_x = \pm\sqrt{-1 \mp i\pi} \text{ fm} \quad (\text{or, equivalently}, c = i \text{ fm}).$$

As explained above, the angular distributions for the two phase shift functions coincide at small momentum transfers. For larger momentum transfers, however, the two phase shift functions lead to quite different cross sections. The cross section corresponding to χ_l acquires a more slowly varying, asymptotically constant, period of oscillation,

$$\Delta q = \frac{\pi}{2\,\mathrm{Re}\sqrt{c^2 - i\pi\beta^2}}, \tag{7.64}$$

and similarly a more slowly varying, asymptotically constant, slope of the envelope,

$$2\left|\,\mathrm{Im}\sqrt{c^2 - i\pi\beta^2}\right|. \tag{7.65}$$

These different ways in which the period of oscillation, and the slope of the envelope change with momentum transfer for the two cases, are related to the variation of the stationary points with momentum transfer. The stationary points for the Gaussian and the logarithmic phase shift functions are compared in the small graphs inserted in Fig. 7.8. For the logarithmic phase shift function the stationary points approach the singularities in the complex b_x-plane, located at $\pm 1.466 - 1.072i$ fm and $\pm 1.072 - 1.466i$ fm, respectively, for the two sets of parameters. Thus, from Eqs. (7.64) and (7.65) it follows that the period of oscillation will be shorter by about 30% and the slope of the cross section envelope will also be smaller by about 30% than in the second case. This may be seen from a comparison of the cross

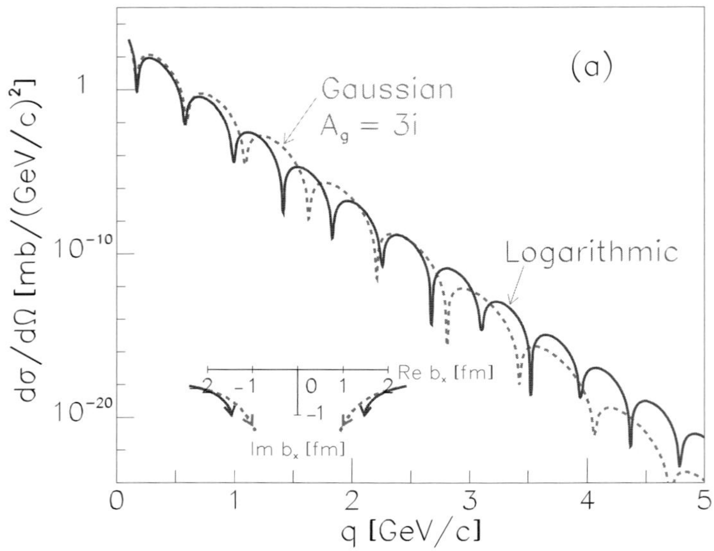

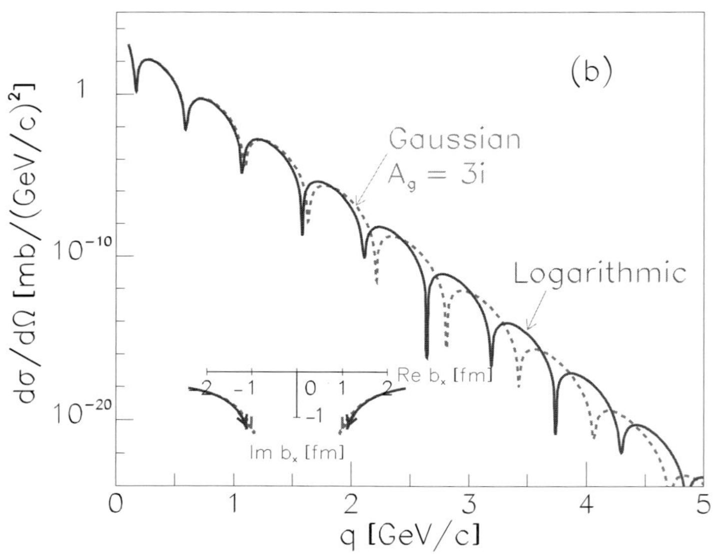

Figure 7.8 Comparisons of the differential cross sections resulting from two phase shift functions having equal edges, $\chi_g(b) = A_g \exp(-b^2/\beta^2)$ and $\chi_l(b) = A_\ell \log\{1 + \exp[(c^2 - b^2)/\beta^2]\}$. The parameters are chosen such that $A_g = A_\ell \exp(c^2/\beta^2) = 3i$, and $\beta = 1$ fm. Two positions are considered for the logarithmic singularity, (a) $b_x = \pm\sqrt{1 \mp i\pi}$ fm (or $c = 1$ fm), (b) $b_x = \pm\sqrt{-1 \mp i\pi}$ fm (or $c = i$ fm). The differential cross sections and the trajectories of stationary points corresponding to χ_g are shown as dashed curves, whereas those corresponding to χ_l are shown as solid curves.

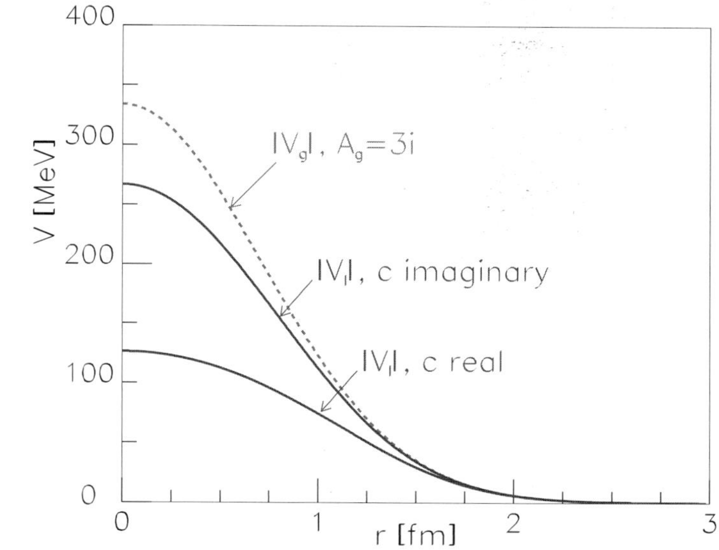

Figure 7.9 Moduli of the potentials corresponding to the Gaussian and logarithmic phase shift functions considered in Fig. 7.8.

sections given in Fig. 7.8(a) and (b), using the cross section for the Gaussian phase shift as a common reference.

It is also instructive to compare the potentials $V_g(r)$ and $V_l(r)$ that correspond to $X_g(b_x)$ and $X_l(b_x)$, respectively. The potential $V_g(r)$ was given in Eq. (7.55), which we rewrite as

$$V_g(r) = -\frac{vA_g}{\sqrt{\pi}\beta}\, e^{-r^2/\beta^2}, \tag{7.66}$$

whereas $V_l(r)$ may be determined from the Abel equation [cf. Eq. (2.6a)],

$$V_l(r) = -\frac{2vA_\ell}{\pi\beta^2}\int_0^\infty \mathrm{d}\xi \left[1 + e^{(r^2-c^2+\xi^2)/\beta^2}\right]^{-1}. \tag{7.67}$$

These potentials are graphed in Fig. 7.9, for the sets of parameters presented in Fig. 7.8. They are seen to be identical at large values of r, but to differ considerably for small r.

7.5 "Fermi–Gaussian" Phase Shift Function

The third phase shift function we consider with a Gaussian edge is the "Fermi–Gaussian" of Eq. (7.3),

$$\chi_{\mathrm{Fg}}(b_x, b_y) = A_{\mathrm{Fg}}/\{1 + \exp[(b^2 - c^2)/\beta^2]\}. \tag{7.68}$$

Like the logarithmic one, $X_{\mathrm{Fg}}(b_x)$ also has singularities at

$$b_x^2 = \pm\sqrt{c^2 \mp i\pi\beta^2}. \tag{7.69}$$

The equation for the stationary points,

$$-\frac{2b_x}{\beta^2}\frac{A_{\mathrm{Fg}}E}{(1+E)^2} = q, \tag{7.70}$$

with

$$E \equiv \exp[(b_x^2 - c^2)/\beta^2], \tag{7.71}$$

may be solved by the method outlined in Appendix B. As $q \to \infty$, the solutions will approach the singularities, given by Eq. (7.69). But X_{Fg} has pole singularities, whereas X_l has logarithmic singularities; as a consequence the stationary points approach the singularities at different rates in the two cases.

In order to discuss the asymptotic properties of the differential cross section, let us write the solution to Eq. (7.70) as

$$b_x^2 = c^2 + \beta^2[(2n+1)i\pi + \log(\bar{E})], \tag{7.72}$$

with

$$\bar{E} = 1 + \frac{A_{\mathrm{Fg}}b_x}{q\beta^2} + \left[\frac{2A_{\mathrm{Fg}}b_x}{q\beta^2} + \left(\frac{A_{\mathrm{Fg}}b_x}{q\beta^2}\right)^2\right]^{1/2}. \tag{7.73}$$

Asymptotically, i.e., for small $|A_{\mathrm{Fg}}|/q\beta$, the solution may be approximated as

$$b_x = \pm\left\{c^2 \mp \beta^2\left[i\pi - \left(\frac{2A_{\mathrm{Fg}}b_x}{q\beta^2}\right)^{1/2}\right]\right\}^{1/2}, \tag{7.74}$$

where we have chosen the signs such as to give the relevant branches of the multivalued expressions Eqs. (7.72 and 7.73). If we now denote the relevant singular points in the third and fourth quadrants as

$$\begin{aligned} b_x^{(s,\alpha)} &= -\sqrt{c^2 + i\pi\beta^2}, \\ b_x^{(s,\beta)} &= \sqrt{c^2 - i\pi\beta^2}, \end{aligned} \tag{7.75}$$

then we have asymptotically

$$\begin{aligned} b_x^{(\alpha)} &= b_x^{(s,\alpha)} + i\beta\left(\frac{A_{\mathrm{Fg}}}{-2qb_x^{(s,\alpha)}}\right)^{1/2}, \\ b_x^{(\beta)} &= b_x^{(s,\beta)} + \beta\left(\frac{A_{\mathrm{Fg}}}{2qb_x^{(s,\beta)}}\right)^{1/2}. \end{aligned} \tag{7.76}$$

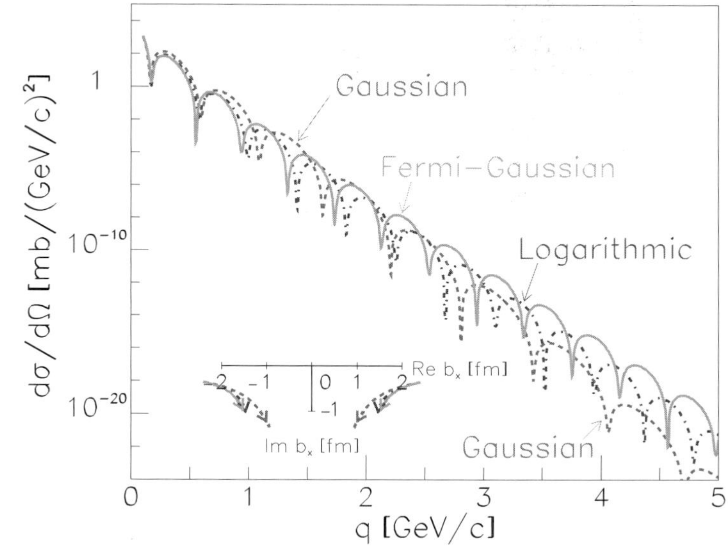

Figure 7.10 Differential cross sections resulting from three phase shift functions that coincide at large impact parameters, but have different analytic properties, $\chi_{\mathrm{g}}(b) = A_{\mathrm{g}} \exp(-b^2/\beta^2)$ (dashed), $\chi_l(b) = A_\ell \log\{1+\exp[(c^2-b^2)/\beta^2]\}$ (dash-dotted), and $\chi_{\mathrm{Fg}}(b) = A_{\mathrm{Fg}}/\{1+\exp[(b^2-c^2)/\beta^2]\}$ (solid). The parameters satisfy Eq. (7.78). Further, $A_{\mathrm{g}} = 3i$, $\beta = 1$ fm, and $c = 1$ fm (as in Fig. 7.8(a)). The trajectories of stationary points are given in the insert.

For the logarithmic phase shift function, the approach to the singularity is faster. A similar expansion of Eq. (7.59) furnishes the asymptotic relations

$$b_{\alpha x} = b_x^{(s,\alpha)} - \frac{A_\ell}{q},$$

$$b_{\beta x} = b_x^{(s,\beta)} + \frac{A_\ell}{q}. \tag{7.77}$$

Comparing Eqs. (7.76) and (7.77), we also note that the singularities are approached from different directions for the two phase shift functions. This has some bearing on the relative rates at which the envelope and the period of the differential cross section change as functions of momentum transfer.

We show in Fig. 7.10 a comparison of the differential cross sections resulting from X_{g}, X_l, and X_{Fg}, for pure absorption, and with relative strengths chosen such that the phase shift functions coincide at large impact parameters,

$$A_{\mathrm{Fg}}\, e^{c^2/\beta^2} = A_\ell\, e^{c^2/\beta^2} = A_{\mathrm{g}}, \tag{7.78}$$

compare Eq. (7.63). With this condition satisfied, it is easy to show that the differential cross section resulting from X_{Fg} at small momentum transfers coincides with

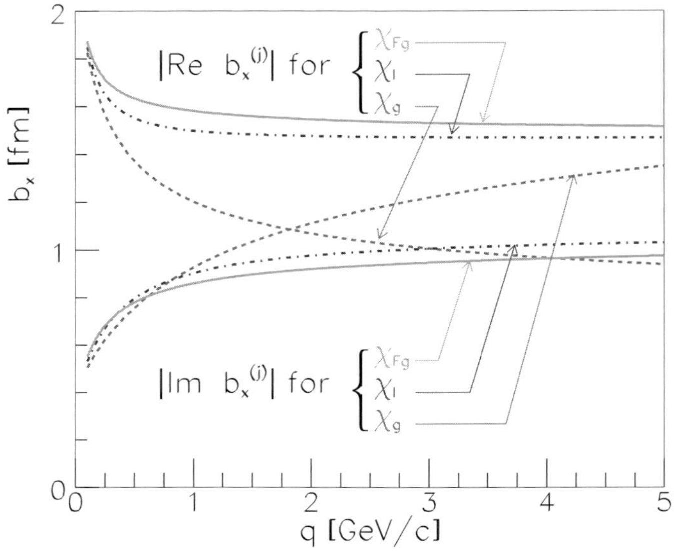

Figure 7.11 Variation of stationary phase coordinates b_{jx} with momentum transfers, for the three phase shift functions considered in Fig. 7.10. The moduli of the real and imaginary parts are shown separately.

those corresponding to the Gaussian and logarithmic phase shift functions. This is seen to be the case in Fig. 7.10, where we have taken the same parameters as in Fig. 7.8(a), i.e., $A_g = 3i$, $\beta = c = 1$ fm.

As the momentum transfer increases, the period of the differential cross section for X_{Fg} is seen to be even more stable than it was for the logarithmic phase shift function. This is largely due to the slower approach of the stationary points to the singularities, as given by Eqs. (7.76) and (7.77) for the two cases. For the parameters considered in Fig. 7.10, the actual q-dependence of the stationary points is shown in Fig. 7.11 for these three phase shift functions.

The differential cross sections corresponding to X_l and X_{Fg} both appear at large momentum transfers to oscillate with nearly constant periods, but to be shifted by half a period with respect to one another. One may determine the way in which the periods approach constancy by evaluating, in addition to the stationary points, the phase shift functions at the stationary points.

At the stationary points, by using $\bar{E}$ defined by Eq. (7.73) we may write

$$X_{Fg}(b_{jx}) = -\frac{q\beta^2}{2b_{jx}}(1 - \bar{E}^{-1}), \tag{7.79}$$

and

$$X''_{Fg}(b_{jx}) = -\frac{2qb_{jx}}{\beta^2}\left[\frac{\bar{E}+1}{\bar{E}-1} - \frac{\beta^2}{2b_{jx}^2}\right]. \tag{7.80}$$

As $q \to 0$, these expressions are seen to become identical to those for the Gaussian phase shift function, as is required for the cross sections to coincide.

Asymptotically, it follows that

$$X_{\mathrm{Fg}}(b_{jx}) \simeq -\left(\frac{q\beta^2 A_{\mathrm{Fg}}}{2b_{jx}}\right)^{1/2}, \tag{7.81}$$

whereas [compare Eq. (7.61)]

$$X_l(b_{jx}) \simeq -A_\ell \log\left(\frac{q\beta^2}{2b_{jx}A_\ell}\right). \tag{7.82}$$

We may combine these results with those for the stationary points, Eqs. (7.76) and (7.77), and obtain thus for the case of pure absorption the dominant q-dependences of the periods:

For X_l:

$$\Delta q \simeq \frac{\pi}{2\,\mathrm{Re}\left[b_x^{(s,\beta)} + \dfrac{A_\ell}{q}\log\left(\dfrac{\beta}{2b_x^{(s,\beta)}A_\ell}\right)\right]}, \tag{7.83}$$

for X_{Fg}:

$$\Delta q \simeq \frac{\pi}{2\,\mathrm{Re}\left[b_x^{(s,\beta)} + \beta\left(\dfrac{2A_{\mathrm{Fg}}}{qb_x^{(s,\beta)}}\right)^{1/2}\right]}. \tag{7.84}$$

(A_ℓ and A_{Fg} are here assumed to be purely imaginary.)

Since the periods of the differential cross sections for the logarithmic and the "Fermi–Gaussian" phase shifts approach their asymptotically constant values at different rates, the phase of one oscillatory pattern will drift with respect to that of the other. At the largest momentum transfers considered in Fig. 7.10, they are shifted by half a period. This shift would increase if the cross sections were followed out to still larger momentum transfers.

For these regularly oscillating differential cross sections, it is useful to introduce the concept of envelopes as the smooth curves to which the differential cross section curves are tangential (analogous to the straight lines in Fig. 1.1). If we denote the exponential slopes of these envelopes by $S(q)$, then the differential cross sections fall as $\exp[-S(q)q]$. The slopes of these envelopes of the cross sections may be determined in a fashion similar to that of the periods discussed above. We thus find:

For X_l:

$$S(q) \simeq 2\left|\,\mathrm{Im}\left\{b_x^{(s,\beta)} + \frac{A_\ell}{q}\left[1 + \log\left(\frac{q\beta^2}{2b_x^{(s,\beta)}A_\ell}\right)\right]\right\}\right|, \tag{7.85}$$

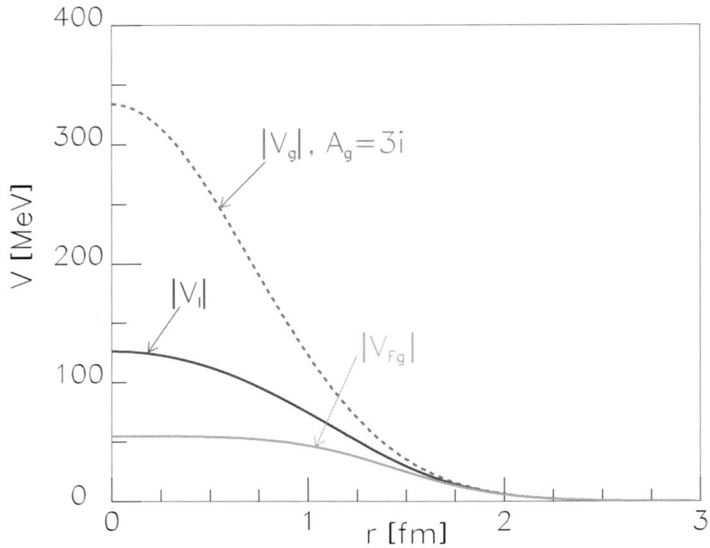

Figure 7.12 Moduli of the potentials corresponding to the Gaussian, logarithmic, and "Fermi–Gaussian" phase shift functions considered in Fig. 7.10.

for $X_{\rm Fg}$:

$$S(q) \simeq 2\left| \mathrm{Im}\left\{ b_x^{(s,\beta)} + \beta\left(\frac{2A_{\rm Fg}}{q b_x^{(s,\beta)}}\right)^{1/2} \right\}\right|, \tag{7.86}$$

(for purely imaginary A_ℓ and $A_{\rm Fg}$). For the logarithmic phase shift function the asymptotic value of the slope,

$$S(q) \simeq 2|\,\mathrm{Im}\, b_x^{(s,\beta)}| = 2\,\mathrm{Im}\sqrt{c^2 + i\pi\beta^2}, \tag{7.87}$$

is approached more rapidly than for the "Fermi–Gaussian". We note that the correction term for the "Fermi–Gaussian", proportional to $q^{-1/2}$, tends to reduce the slope. This explains why, at large momentum transfers in Fig. 7.10, the envelope of the differential cross section for the "Fermi–Gaussian" phase shift falls less rapidly than that for the logarithmic one.

Finally, in Fig. 7.12, we compare the three potentials that correspond to the differential cross sections of Fig. 7.10. The potential $V_{\rm Fg}(r)$ that corresponds to $\chi_{\rm Fg}(b)$ has been determined from the Abel equation,

$$V_{\rm Fg}(r) = -\frac{2vA_{\rm Fg}}{\pi\beta^2}\int_0^\infty \mathrm{d}\xi\, \frac{\exp[(r^2 - c^2 + \xi^2)/\beta^2]}{\{1 + \exp[(r^2 - c^2 + \xi^2)/\beta^2]\}^2}, \tag{7.88}$$

while $V_g(r)$ and $V_l(r)$ are as given in Fig. 7.8. Since the parameters considered are such that the three phase shift functions agree at large b, the potentials will similarly agree at large radii, whereas they differ considerably in magnitude at small r.

We have seen above that the asymptotic theory enables us to determine important features of the diffractive pattern from properties of the phase shift function. In particular, we have seen that the analytic properties of the phase shift function (or of the potential), which are not apparent from simple graphs of these quantities, are of great importance at large momentum transfers.

8

Phase Shift Functions with Exponential Edge

The surface densities of heavy nuclei can be well represented by exponential functions. In fact, the function familiar from its role in the thermal Fermi distribution is widely used as an approximation in the description of densities of heavy nuclei. Such (Woods–Saxon) densities lead to phase shift functions that also have exponential edges. The present chapter is devoted to a survey of some properties of phase shift functions that have such an exponential edge. We have chosen to consider three examples,

$$\begin{aligned}\chi_{\exp}(b_x, b_y) &= A_{\exp}\, e^{-b/\beta} \\ &= A_{\exp} \exp\Bigl(-\sqrt{b_x^2 + b_y^2}/\beta\Bigr),\end{aligned} \tag{8.1}$$

$$\begin{aligned}\chi_{\cosh}(b_x, b_y) &= \frac{A_{\cosh}}{\cosh(b/\beta)} \\ &= \frac{A_{\cosh}}{\cosh\bigl(\sqrt{b_x^2 + b_y^2}/\beta\bigr)},\end{aligned} \tag{8.2}$$

$$\begin{aligned}\chi_{\mathrm{F}}(b_x, b_y) &= \frac{A_{\mathrm{F}}}{1 + \exp[(b - c)/\beta]} \\ &= \frac{A_{\mathrm{F}}}{1 + \exp\Bigl[\Bigl(\sqrt{b_x^2 + b_y^2} - c\Bigr)/\beta\Bigr]},\end{aligned} \tag{8.3}$$

which we shall refer to as the exponential, the inverse cosh, and the "Fermi" phase shift function, respectively.

In analogy with the comparison of phase shift functions with a Gaussian edge in Fig. 7.1, we also compare the three phase shift functions with an exponential edge in Fig. 8.1. For this purpose, we choose the relative normalizations

$$A_{\cosh} = A_{\exp}/2, \quad A_{\mathrm{F}} = A_{\exp}/\exp(c/\beta), \tag{8.4}$$

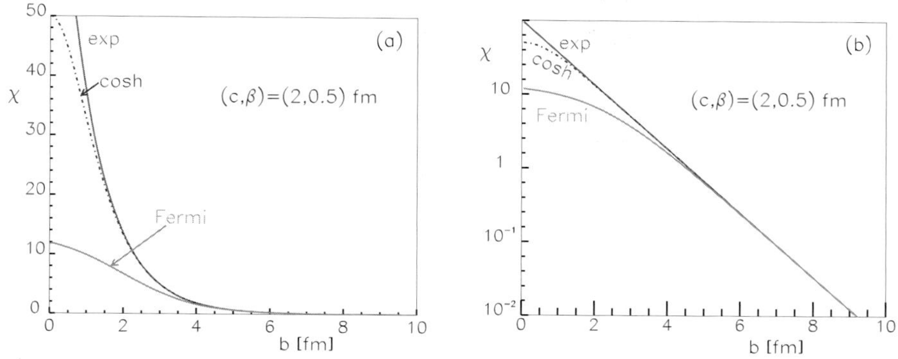

Figure 8.1 Phase shift functions Eqs. (7.1–7.3) with an exponential edge. The same three phase shift functions are shown both in linear (a) and logarithmic (b) representations.

and use the illustrative values $A_{\rm exp} = 100$, $c = 2$ fm, and $\beta = 0.5$ fm. At first sight (Fig. 8.1(a), where they are plotted linearly) they are rather different. However, in the case of strong absorption, it is the edge region which is important. This region is shown on a logarithmic scale, in Fig. 8.1(b). With the normalization of Eq. (8.4), the edge regions for these three phase shift functions are essentially the same in magnitude, as will be the envelopes of their patterns of oscillations.

Because of cylindrical symmetry, in the asymptotic approach we only need these phase shift functions for $b_y = 0$, and therefore define

$$\begin{aligned} X_{\rm exp}(b_x) &= A_{\rm exp}\, e^{-b_x/\beta}, \qquad\qquad \operatorname{Re} b_x > 0, \\ X_{\rm exp}(-b_x) &= X_{\rm exp}(b_x), \end{aligned} \tag{8.5}$$

$$X_{\cosh}(b_x) = \frac{A_{\cosh}}{\cosh(b_x/\beta)}, \tag{8.6}$$

$$\begin{aligned} X_{\rm F}(b_x) &= \frac{A_{\rm F}}{1+\exp[(b_x-c)/\beta]}, \qquad \operatorname{Re} b_x > 0, \\ X_{\rm F}(-b_x) &= X_{\rm F}(b_x). \end{aligned} \tag{8.7}$$

Since the pure exponential and the "Fermi" phase shift functions are not analytic at the origin, we need to give explicit definitions for $\operatorname{Re} b_x < 0$. Eqs. (8.5) and (8.7) correctly represent continuations of phase shift functions as even-valued in b_x. We can use these continuations as long as the stationary points do not get close to the origin.

In the case of the "Fermi" phase shift function, one can easily construct variations of Eq. (8.3) that are even and analytic at the origin. These include

$$\tilde{\chi}_{\mathrm{F}}(b_x, b_y) = \frac{A}{\{1+\exp[(b-c)/\beta]\}\{1+\exp[(-b-c)/\beta]\}}, \tag{8.8}$$

and

$$\bar{\chi}_{\mathrm{F}}(b_x, b_y) = A\{[1+\exp[(-b-c)/\beta]]^{-1} + [1+\exp[(b-c)/\beta]]^{-1} - 1\}. \tag{8.9}$$

For large values of $|\boldsymbol{b}|/\beta$, i.e., at the nuclear surface, the three forms, Eqs. (8.1–8.3) have the same exponential behavior. As we shall see, for suitable choices of the parameters, and for small momentum transfers q, they therefore all lead to similar cross sections. However, their analytic properties in the complex b_x-plane are different (even disregarding the subtleties along the line $\mathrm{Re}\, b_x = 0$) and at sufficiently large momentum transfers the differential cross sections will therefore differ.

8.1 Evenness

Since we integrate over Cartesian coordinates b_x and b_y rather than cylindrical ones, we need to define the integrand carefully for negative values of b_x and b_y. The "Fermi" phase shift function Eq. (8.3) is for fixed b_y written as an even function of b_x. After integrating over b_y and evaluating the integrand for $b_y = 0$, the evenness is lost. The construction Eq. (8.7) is explicitly even in b_x, but it is not analytic at the origin. In fact, all odd derivatives are non-vanishing and opposite on the two sides. For the stationary-phase integration, this does not pose any problems, since the integrand is exponentially small near $b_x = 0$ and that region does not give any numerically significant contribution. The same is true for the exponential phase shift function, given by Eqs. (8.1) and (8.5).

For the direct numerical evaluation of the diffraction integral Eq. (7.29) in cylindrical coordinates, however, the distinction between Eqs. (8.3) and (8.8) is important at large momentum transfers. Without the evenness imposed by Eq. (8.8), the scattering amplitude would start to exhibit noise-like behavior at large momentum transfers.

8.2 Exponential Phase Shift Function

Consider first $X_{\exp}(b_x)$. The stationary points can be determined analytically as

$$b_{\alpha x} = \beta\left[-\log\left(\frac{|A_{\exp}|}{q\beta}\right) - i\varphi_A\right], \tag{8.10a}$$

$$b_{\beta x} = \beta\left[\log\left(\frac{|A_{\exp}|}{q\beta}\right) - i(\pi - \varphi_A)\right], \tag{8.10b}$$

where

$$A_{\text{exp}} \equiv |A_{\text{exp}}| e^{i\varphi_A}. \tag{8.11}$$

Now, using the values of X_{exp} and X''_{exp} at the stationary points,

$$\begin{aligned} X_{\text{exp}}(b_{\alpha x}) &= -X_{\text{exp}}(b_{\beta x}) = q\beta, \\ X''_{\text{exp}}(b_{\alpha x}) &= -X''_{\text{exp}}(b_{\beta x}) = q/\beta, \end{aligned} \tag{8.12}$$

we obtain simple expressions for the corresponding amplitudes,

$$F = F(b_{\alpha x}) + F(b_{\beta x}), \tag{8.13}$$

with

$$F(b_{\alpha x}) = \frac{k}{iq}\sqrt{\beta b_{\alpha x}}\, e^{-iq(b_{\alpha x}-\beta)}, \tag{8.14a}$$

$$F(b_{\beta x}) = \frac{k}{iq}\sqrt{-\beta b_{\beta x}}\, e^{-iq(b_{\beta x}+\beta)}. \tag{8.14b}$$

The moduli of each of these will obviously fall as

$$|F(b_{\alpha x})| \sim e^{-q\beta\varphi_A}, \tag{8.15a}$$

$$|F(b_{\beta x})| \sim e^{-q\beta(\pi-\varphi_A)}. \tag{8.15b}$$

The different rates at which these fall off are thus simply given by the phase φ_A, and are equal for $\varphi_A = \pi/2$. Similarly, the period of oscillation in the cross section is given by

$$\begin{aligned} \Delta q &= \frac{\pi}{|\operatorname{Re}[b_{\alpha x} - b_{\beta x} - 2\beta]|} \\ &= \frac{\pi}{2\left[\log\left(\frac{|A_{\text{exp}}|}{q\beta}\right) + \beta\right]}. \end{aligned} \tag{8.16}$$

Thus, we see that the asymptotic method provides us with simple and explicit formulas for the characteristics of the cross section.

In Fig. 8.2 we display the trajectories of the stationary points for $\varphi_A = 0°$, $30°$, $60°$, and $90°$. The arrows indicate the movement of the points of stationary phase, as the momentum transfer, q, is increased. Note that there is no rainbow singularity associated with $X_{\text{exp}}(b_x)$ for $\varphi_A = 0°$.

The cross sections corresponding to $\varphi_A = 30°$ and $90°$ are shown in Figs. 8.3 and 8.4. Fig. 8.3 shows characteristic damping of the oscillations due to the different imaginary parts of the trajectories. Fig. 8.4 shows oscillations with increasing period as q increases as the two stationary points approach each other. Were the

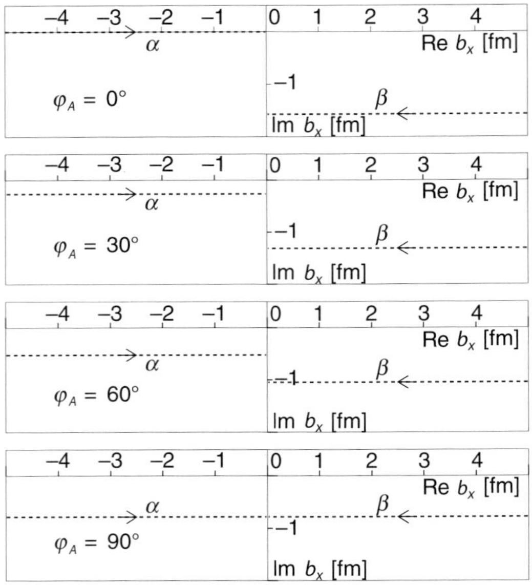

Figure 8.2 Trajectories of stationary points for the exponential phase shift function, $\chi_{\rm exp} = A_{\rm exp}\exp(-b/\beta)$. [Its definition for Cartesian coordinates, $\boldsymbol{b} = (b_x, b_y)$ is given by Eq. (8.1).] These trajectories are shown in the complex b_x-plane, the arrows indicate how they move with increasing q. (At the stationary points, $b_y = 0$.) Here $A_{\rm exp} = |A_{\rm exp}|\exp(i\varphi_A)$; four phases are considered, $\varphi_A = 0°, 30°, 60°$, and $90°$. Further, $\beta = 0.5$ fm. The present simple phase shift function, which is only an approximation to more realistic phase shift functions, has a discontinuous derivative along the line $\mathrm{Re}\, b_x = 0$. That line separates the two trajectories labeled α and β. For the case of pure absorption, the two trajectories satisfy $\mathrm{Im}\, b_{jx} = -\frac{1}{2}\pi\beta = -0.785$ fm.

imaginary parts of the stationary point coordinates the only parameters determining the slope of the envelope of the differential cross section, the envelope would have a constant slope on a logarithmic plot. The more slowly varying factors in the expression for the scattering amplitude are seen, on the contrary, to lead to a slight deviation from this constancy of slope.

8.3 Inverse cosh Phase Shift Function

Next, we consider $X_{\rm cosh}(b_x)$ given by Eq. (8.6). Elementary algebra leads to explicit expressions for the stationary points. These take a simple form if we introduce the abbreviations

$$\lambda = \frac{A_{\rm cosh}}{2q\beta}, \tag{8.17}$$

$$s_\pm = -\lambda \pm i\sqrt{1-\lambda^2}, \tag{8.18}$$

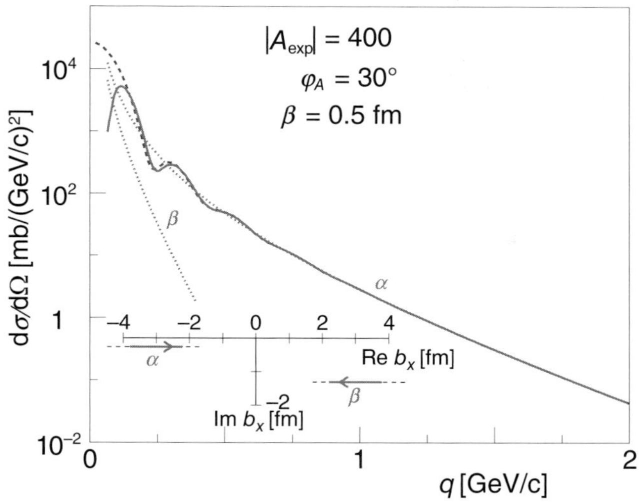

Figure 8.3 Differential cross section for $\chi_{exp}(b) = A_{exp}\exp(-b/\beta)$. Here $A_{exp} = |A_{exp}|\exp(i\varphi_A)$, with $|A_{exp}| = 400$. The phase shift function has a small absorptive part, $\varphi_A = 30°$, and $\beta = 0.5$ fm. The inserts show the trajectories of stationary points for the range in momentum transfer 0.02 GeV/c $\leq q \leq$ 6.0 GeV/c (dotted). The parts of the trajectories that correspond to the cross-section plots, 0.05 GeV/c $\leq q \leq$ 2.0 GeV/c, are indicated as solid curves. (The arrows indicate the sense in which the stationary points move as the momentum transfer increases.) The differential cross sections contributed by each of the stationary points, α and β, are shown as dotted curves. The solid curve shows the resulting cross section, evaluated in the asymptotic approach, whereas the dashed curve shows the result of a numerical evaluation of the diffraction integral, given by Eq. (2.2).

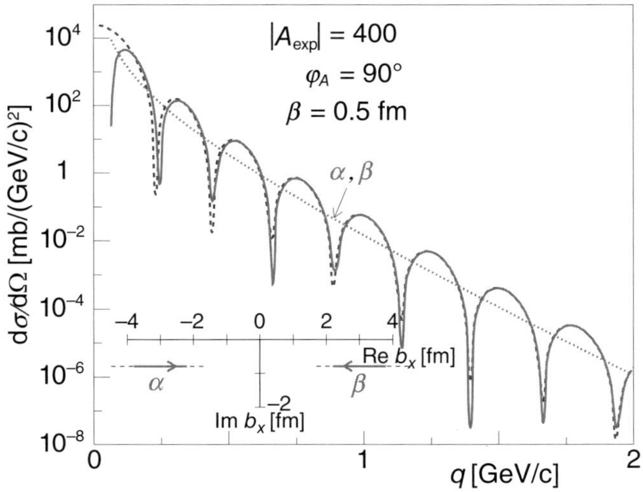

Figure 8.4 Similar to Fig. 8.3, for $\varphi_A = 90°$.

namely

$$b_{jx} = \beta \log\left[s_- \pm \sqrt{s_-^2 + 1}\right], \qquad j = \alpha, \beta, \tag{8.19a}$$

$$b_{jx} = \beta \left\{\log\left[s_+ \pm \sqrt{s_+^2 + 1}\right] - 2\pi i\right\}, \qquad j = \gamma, \delta. \tag{8.19b}$$

Out of the infinite number of periodically repeating solutions, with period $2\pi i \beta$, we have selected those satisfying the inequality $-2\pi\beta < \operatorname{Im} b_{jx} \leq 0$ and through which the path of integration must be passed. For cases with absorption, only $b_{\alpha x}$ and $b_{\beta x}$ are relevant in this way, whereas for $A_{\cosh}$ real and positive (negative), $b_{\gamma x}$ ($b_{\delta x}$) will also be needed. For $\varphi_A = 0°, 30°, 60°$, and $90°$, these stationary points are shown in Fig. 8.5. The case $\varphi_A = 90°$ corresponds to the function discussed in Section 5.1.

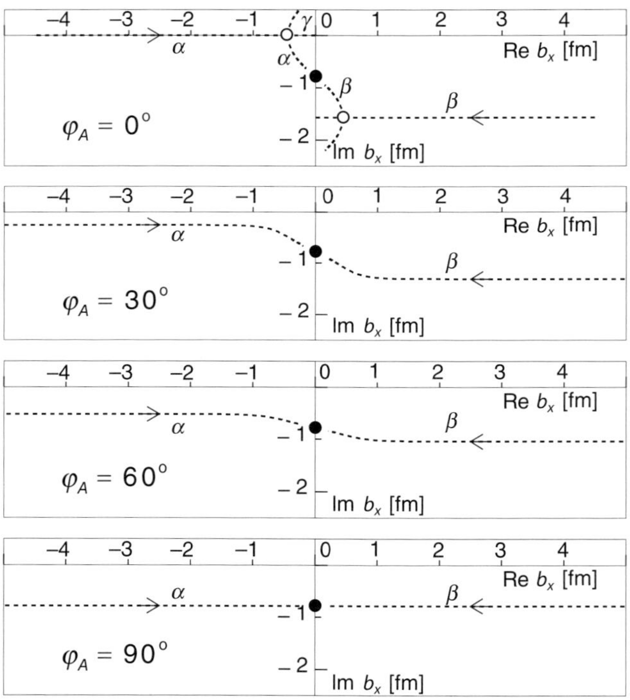

Figure 8.5 Trajectories of stationary points for $\chi_{\cosh}(b) = A_{\cosh}/\cosh(b/\beta)$, with $A_{\cosh} = |A_{\cosh}| \exp(i\varphi_A)$. These trajectories are shown in the complex b_x-plane, where $\boldsymbol{b} = (b_x, b_y)$. (At the stationary point, $b_y = 0$.) Four different phases are considered, $\varphi_A = 0°, 30°, 60°$, and $90°$, and $\beta = 0.5$ fm. The filled circles indicate poles of the phase shift function, whereas the open circles indicate zeros of its second derivative. One of these zeros, the one where for $\varphi_A = 0°$ the trajectories α and γ run together, is a rainbow point. The other zero shown for $\varphi_A = 0°$ (in the fourth quadrant) is not a rainbow point. In fact, neither the trajectory labeled β, nor its counterpart on the other side of the open circle, is encountered along the path of integration for $\varphi_A = 0°$. (Compare the insert in Fig. 8.6.)

Rainbow points can occur where

$$X''_{\cosh}(b_x) = -\frac{A_{\cosh}}{\beta^2}\frac{1-\sinh^2(b_x/\beta)}{\cosh^2(b_x/\beta)} = 0, \tag{8.20}$$

if $X'_{\cosh}(b_x)$ is simultaneously real and positive. These conditions are met for $A_{\cosh}$ real, and then correspond to

$$b_{x,\mathrm{R}} = -\beta\log(\sqrt{2}+1), \quad \text{for } A_{\cosh} > 0, \tag{8.21a}$$

$$b_{x,\mathrm{R}} = \beta\log(\sqrt{2}+1), \quad \text{for } A_{\cosh} < 0. \tag{8.21b}$$

The stationary point trajectories are seen to turn suddenly at right angles at the rainbow point (small circle) in Fig. 8.5. The rainbow singularity occurs for

$$q_{\mathrm{R}} = A_{\cosh}/2\beta. \tag{8.22}$$

In Fig. 8.6 we show the cross section for a real phase shift function with $A_{\cosh} = 10$ and $\beta = 0.5$ fm. A comparison with the exact evaluation of the diffraction

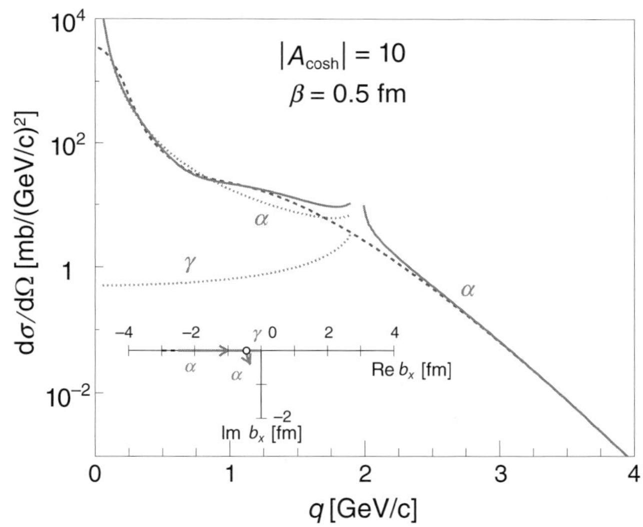

Figure 8.6 Differential cross sections for $\chi_{\cosh}(b) = A_{\cosh}/\cosh(b/\beta)$, with $A_{\cosh} = 10$ and $\beta = 0.5$ fm. The insert shows the trajectories of stationary points for the range in momentum transfers, 0.02 GeV/c $\leq q \leq$ 6.0 GeV/c (dotted). The parts of the trajectories that correspond to the cross-section plot, 0.05 GeV/c $\leq q \leq$ 4.0 GeV/c, are indicated as solid curves. (The arrows indicate the sense in which the stationary points move as the momentum transfer is increased.) As may be seen from a study of maps of the type given in Figs. 5.6 and 11.6, for a real, positive $A_{\cosh}$, the trajectory labeled β in Fig. 8.5 is not encountered along the path of integration. The differential cross sections contributed by each of the stationary points α and γ, are shown as dotted curves. The solid curve shows the resulting cross section, evaluated in the asymptotic approach (the spike represents the rainbow singularity), whereas the dashed curve shows the result of a numerical evaluation of the diffraction integral.

integral shows that the agreement is excellent, apart from small intervals close to $q = 0$ and $q = q_R$. These features are quite analogous to those encountered for the real Gaussian phase shift function. (Compare Fig. 7.3 and the corresponding discussion in Chapter 7.)

Consider for a moment parameters $A_{\cosh}$, β, and a range in q such that

$$|\lambda| \equiv \frac{|A_{\cosh}|}{2q\beta} \gg 1. \tag{8.23}$$

We then have

$$b_{\alpha x} \simeq \beta\left[-\log\left(\frac{2|A_{\cosh}|}{q\beta}\right) - i\varphi_A\right], \tag{8.24a}$$

$$b_{\beta x} \simeq \beta\left[\log\left(\frac{2|A_{\cosh}|}{q\beta}\right) - i(\pi - \varphi_A)\right], \tag{8.24b}$$

which for $2A_{\cosh} = A_{\exp}$ are the same as for the exponential case, Eq. (8.10). In this limit, we also find that $X_{\cosh}(b_{jx})$ and $X''_{\cosh}(b_{jx})$ are equal to $X_{\exp}(b_{jx})$ and $X''_{\exp}(b_{jx})$, respectively. It follows that in the range of momentum transfers satisfying Eq. (8.23), one cannot distinguish between $X_{\exp}(b_x)$ and $X_{\cosh}(b_x)$.

Let us now turn to a discussion of weaker interactions or larger momentum transfers, such that

$$|\lambda| \equiv \frac{|A_{\cosh}|}{2q\beta} \ll 1. \tag{8.25}$$

Neglecting terms of order $|\lambda|$, we find from Eqs. (8.18) and (8.19) that

$$b_{jx} \simeq \beta\left[\pm(-1+i)\left(\frac{|A_{\cosh}|}{2q\beta}\right)^{\frac{1}{2}} e^{i\varphi_A/2} - \frac{i\pi}{2}\right], \quad j = \alpha, \beta. \tag{8.26}$$

Thus

$$\operatorname{Re} b_{jx} \simeq \mp\beta\left(\frac{|A_{\cosh}|}{2q\beta}\right)^{\frac{1}{2}}\left(\cos\frac{\varphi_A}{2} + \sin\frac{\varphi_A}{2}\right), \tag{8.27}$$

and

$$\operatorname{Im} b_{jx} \simeq -\beta\left[\frac{\pi}{2} \mp \left(\frac{|A_{\cosh}|}{2q\beta}\right)^{\frac{1}{2}}\left(\cos\frac{\varphi_A}{2} - \sin\frac{\varphi_A}{2}\right)\right]. \tag{8.28}$$

In this regime, the period of oscillation will increase as

$$\begin{aligned}\Delta q &= \frac{\pi}{\left|\operatorname{Re}\left\{b_{\alpha x} - b_{\beta x} - \frac{1}{q}\left[X(b_{\alpha x}) - X(b_{\beta x})\right]\right\}\right|} \\ &= \frac{\pi}{\beta}\left(\frac{2q\beta}{|A_{\cosh}|}\right)^{\frac{1}{2}} \Big/ \left(\cos\frac{\varphi_A}{2} + \sin\frac{\varphi_A}{2}\right),\end{aligned} \tag{8.29}$$

which is much faster than the logarithmic behavior found for $|\lambda| \gg 1$. Also, the rate at which the modulus of each term, $|F(b_{\alpha x})|$ and $|F(b_{\beta x})|$, decreases [cf. Eq. (8.28)], is different from what it was for $|\lambda| \gg 1$. In particular, both slopes tend to the same value, $-\pi\beta/2$, as $q \to \infty$, independent of φ_A. This is of course dictated by the singularity at $b_x = -i\pi\beta/2$. Their difference vanishes like $q^{-1/2}$.

This type of behavior is illustrated in Fig. 8.7, where we have plotted $\mathrm{d}\sigma/\mathrm{d}\Omega$ corresponding to $|A_{\cosh}| = 5$, $\varphi_A = 75°$ and $90°$, $\beta = 0.5$ fm. As in Chapter 7, we have adopted the value of the momentum as $k = 1.7$ GeV/c, corresponding

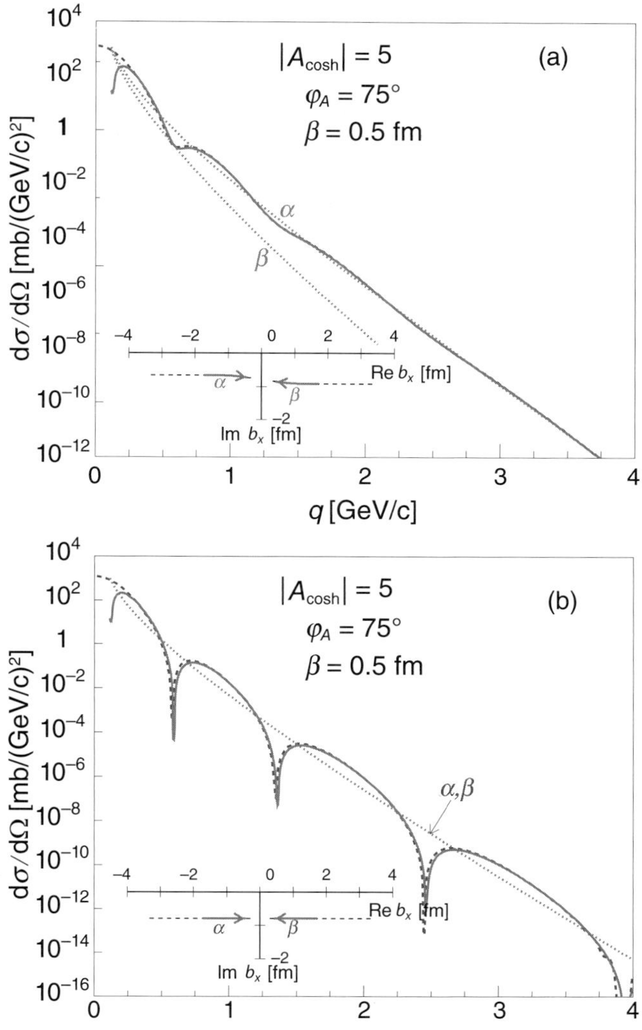

Figure 8.7 Differential cross sections for $\chi_{\cosh}(b) = A_{\cosh}/\cosh(b/\beta)$, with $A_{\cosh} = |A_{\cosh}|\exp(i\varphi_A)$, $|A_{\cosh}| = 5$, and $\beta = 0.5$ fm. Two cases are considered, (a) $\varphi_A = 75°$, and (b) $\varphi_A = 90°$. The individual contributions of the two stationary points (see inserts) are shown as dotted curves, whereas the resulting interference patterns are shown by the solid curves. The dashed curves give the results of numerical evaluations of the diffraction integrals.

to $T_{\rm kin} = 1$ GeV. The value of $|\lambda|$ for this case ranges between 0.99 and 0.25 for q in the interval $1\ \mathrm{GeV/c} \le q \le 4\ \mathrm{GeV/c}$. In Fig. 8.7a the differential cross section for $\varphi_A = 75^\circ$ shows a sequence of damped oscillations of a sort frequently encountered in diffraction patterns. The oscillations are of course produced by the interference of the amplitudes contributed by the stationary points labeled α and β. The damping of the oscillations occurs because the two trajectories yield stationary points which have different imaginary parts. The scattered intensity for large q is given almost entirely by the squared modulus of the amplitude corresponding to $b_{\alpha x}$. In Fig. 8.7b, corresponding to a purely imaginary interaction, $\varphi_A = 90^\circ$, the oscillations are undamped, since the two trajectories have equal imaginary parts. In both Figs. 8.7a and 8.7b, a lengthening of the period of oscillation is evident as q increases, since the distance between the stationary points decreases. Both graphs show excellent agreement between the asymptotic theory and the exact evaluation of the diffraction integral.

8.4 "Fermi" Phase Shift Function

The third example of a phase shift function with exponential edge is the "Fermi" form, Eq. (8.7). This differs from the previous two by possessing one more shape parameter, the radius c. As in the two previous cases, the stationary point can once more be determined analytically. As an abbreviation, let us introduce

$$\lambda \equiv \frac{A_{\rm F}}{2q\beta}. \tag{8.30}$$

The stationary points are then given in terms of c, β, and λ for $\varphi_A > 0$ by

$$b_{\alpha x} = -c - i\pi\beta - \beta \log\left[1 - \lambda + \sqrt{(1-\lambda)^2 - 1}\right], \tag{8.31a}$$

$$b_{\beta x} = c - i\pi\beta + \beta \log\left[1 + \lambda + \sqrt{(1+\lambda)^2 - 1}\right]. \tag{8.31b}$$

Similarly, for $\varphi_A = 0^\circ$, we find

$$b_{\alpha x} = -c - i\pi\beta - \beta \log\left[1 - \lambda - i\sqrt{2\lambda - \lambda^2}\right], \tag{8.32a}$$

$$b_{\beta x} = c - i\pi\beta + \beta \log\left[1 + \lambda + \sqrt{2\lambda - \lambda^2}\right], \tag{8.32b}$$

$$b_{\gamma x} = -c + i\pi\beta - \beta \log\left[1 - \lambda + i\sqrt{2\lambda - \lambda^2}\right]. \tag{8.32c}$$

In these expressions, as well as in Eq. (8.31), the signs in front of the square roots correspond to the solutions that are relevant for $|\lambda| \to 0$. For large values of $|\lambda|$,

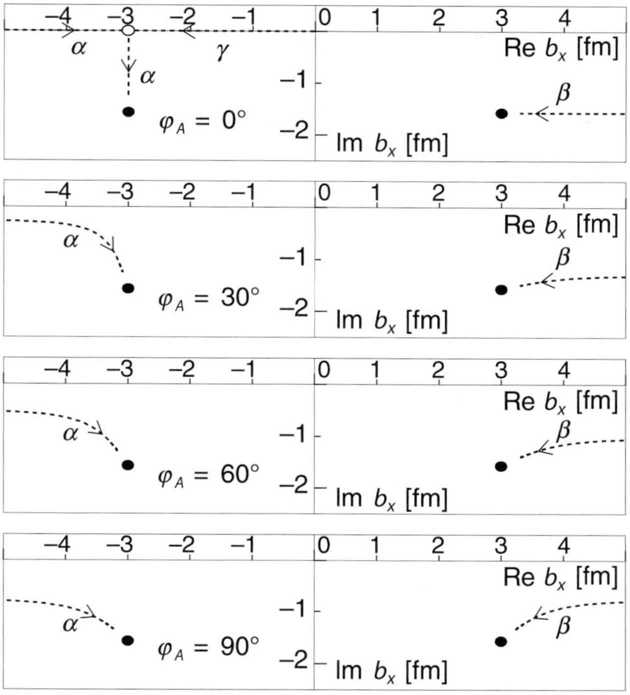

Figure 8.8 Trajectories of stationary points for the "Fermi" phase shift function, $\chi_{\rm F}(b) = A_{\rm F}/\{1+\exp[(b-c)/\beta]\}$. These trajectories are shown in the complex b_x-plane, where $\boldsymbol{b} = (b_x, b_y)$. (At the stationary points, $b_y = 0$.) Again four different phases are considered, $\varphi = 0^\circ, 30^\circ, 60^\circ$, and 90°. The radius and surface thickness parameters are: $c = 3$ fm and $\beta = 0.5$ fm. The filled circles indicate poles of the phase shift function (at $b_x = \pm c - i\pi\beta$), and the open circle (for $\varphi = 0^\circ$) indicates a rainbow point (at $b_x = -c$).

these square roots, as well as the logarithms, must be continued in an analytic fashion, starting with $|\lambda| < 1$.

Asymptotically, as $q \to \infty$, $|\lambda| \to 0$ and

$$b_{\alpha\,x} \simeq -c - i\pi\beta + i\beta e^{i\varphi_A/2}\left(\frac{|A_{\rm F}|}{q\beta}\right)^{1/2}, \tag{8.33a}$$

$$b_{\beta\,x} \simeq c - i\pi\beta + \beta e^{i\varphi_A/2}\left(\frac{|A_{\rm F}|}{q\beta}\right)^{1/2}, \tag{8.33b}$$

i.e., they approach the poles of $X'_{\rm F}(b_x)$.

Trajectories of the stationary points corresponding to $c = 3$ fm, $\beta = 0.5$ fm, and $\varphi_A = 0^\circ$, 30°, 60°, and 90° are given in Fig. 8.8. One may note that, as $q \to \infty$, the poles are approached from different directions, depending on φ_A [cf. Eq. (8.33)].

For small momentum transfers ($|\lambda| \gg 1$), the stationary points are well approximated by

$$b_{\alpha x} = -c - \beta \log\left(\frac{|A_{\rm F}|}{q\beta}\right) - i\beta\varphi_A, \tag{8.34a}$$

$$b_{\beta x} = c + \beta \log\left(\frac{|A_{\rm F}|}{q\beta}\right) - i\beta(\pi - \varphi_A). \tag{8.34b}$$

In this limit, they will coincide with those for $X_{\rm exp}$ [cf. Eq. (8.10)] provided

$$A_{\rm F} = A_{\rm exp}\, e^{-c/\beta}. \tag{8.35}$$

At the stationary point, the phase shift function and its second derivative may be written as

$$X_{\rm F}(b_{\alpha x}) = \tfrac{1}{2}A_{\rm F} - i[A_{\rm F}\, q\beta - \tfrac{1}{4}{A_{\rm F}}^2]^{1/2}, \tag{8.36a}$$

$$X_{\rm F}(b_{\beta x}) = \tfrac{1}{2}A_{\rm F} - [A_{\rm F}\, q\beta + \tfrac{1}{4}{A_{\rm F}}^2]^{1/2}, \tag{8.36b}$$

and

$$X''_{\rm F}(b_{\alpha x}) = \frac{q}{\beta}\left[1 - 2\frac{X_{\rm F}(b_{\alpha x})}{A_{\rm F}}\right], \tag{8.37a}$$

$$X''_{\rm F}(b_{\beta x}) = -\frac{q}{\beta}\left[1 - 2\frac{X_{\rm F}(b_{\beta x})}{A_{\rm F}}\right]. \tag{8.37b}$$

It is now easily seen that when $q\beta/|A_{\rm F}| \ll 1$, and if Eq. (8.35) is fulfilled, the cross section will be the same as for $X_{\rm exp}$. This is yet another example of how the asymptotic method allows us to determine quantitatively when phase shift functions with similar edges yield cross sections which are quite different in form. In the present examples, such differences indeed occur for $q\beta/|A_{\rm F}| \gtrsim 1$ and are illustrated in Fig. 8.12.

For large values of $q\beta/|A_{\rm F}|$, it follows from Eq. (8.36) that

$$X_{\rm F}(b_{\alpha x}) \simeq -i e^{i\varphi_A/2}[q\beta|A_{\rm F}|]^{1/2}, \tag{8.38a}$$

$$X_{\rm F}(b_{\beta x}) \simeq -e^{i\varphi_A/2}[q\beta|A_{\rm F}|]^{1/2}. \tag{8.38b}$$

Thus, the modulus of each term will fall for large q in a nearly exponential fashion,

$$|F(b_{\alpha x})| \sim \exp\left\{-q\beta\left[\pi - 2\cos\left(\frac{\varphi_A}{2}\right)\left(\frac{|A_{\rm F}|}{q\beta}\right)^{1/2}\right]\right\}, \tag{8.39a}$$

$$|F(b_{\beta x})| \sim \exp\left\{-q\beta\left[\pi - 2\sin\left(\frac{\varphi_A}{2}\right)\left(\frac{|A_{\rm F}|}{q\beta}\right)^{1/2}\right]\right\}, \tag{8.39b}$$

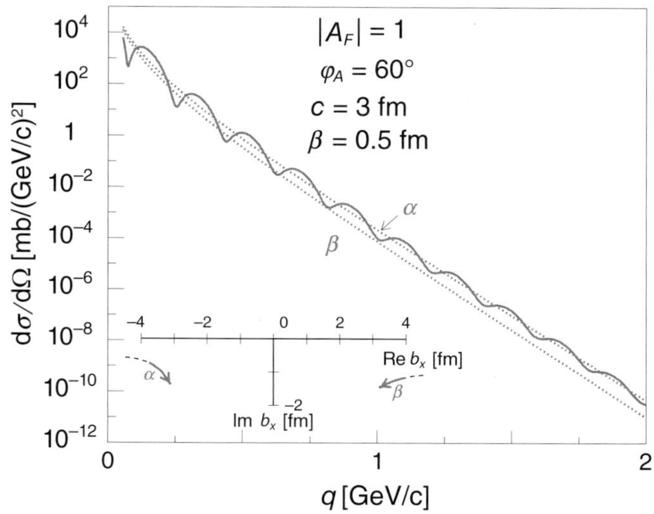

Figure 8.9 Differential cross section for the "Fermi" phase shift function, $\chi_F(b) = A_F/\{1 + \exp[(b - c)/\beta]\}$, with $A_F = |A_F| \exp(i\varphi_A)$. Here $|A_F| = 1$, $c = 3$ fm, and $\beta = 0.5$ fm. A significant degree of absorption is considered, $\varphi_A = 60°$. The inserts show the trajectories of stationary points for the range in momentum transfer, 0.02 GeV/c $\leq q \leq$ 6.0 GeV/c (dotted). The parts of the trajectories that correspond to the cross section plots, 0.05 GeV/c $\leq q \leq$ 2.0 GeV/c, are indicated as solid curves. The differential cross sections contributed by each of the stationary points, α and β, are indicated by dotted curves. The solid curve shows the resulting cross section, evaluated in the asymptotic approach, whereas the dashed curve shows the result of a numerical evaluation of the diffraction integral.

and the period of oscillation will approach its constant asymptotic value, $\pi/2c$ in a similar fashion

$$\Delta q \simeq \frac{\pi}{2\left[c + \beta\left(\cos\frac{\varphi_A}{2} + \sin\frac{\varphi_A}{2}\right)\left(\frac{|A_F|}{q\beta}\right)^{1/2}\right]}. \tag{8.40}$$

We note that the asymptotic period is approached from below.

In Figs. 8.9 and 8.10 we show cross sections corresponding to X_F for $|A_F| = 1$, $c = 3$ fm, and $\beta = 0.5$ fm. In Fig. 8.9 $\varphi_A = 60°$ while in Fig. 8.10 $\varphi_A = 90°$ and X_F is purely imaginary. For the latter case the choice of parameters is roughly appropriate to proton–calcium scattering at 1 GeV. These graphs show two features that are frequently noted as approximate qualities of scattering by heavy nuclei. Their envelopes have nearly constant logarithmic slopes and their periods of oscillation are nearly constant as well. The reason for this behavior of the calculated cross sections lies in the pole singularity of the function X_F given by Eq. (8.1). As the momentum transfer q increases to larger values, the stationary points labeled α and β approach the poles of X_F [see Eq. (8.33)]. In fact, they approach the poles

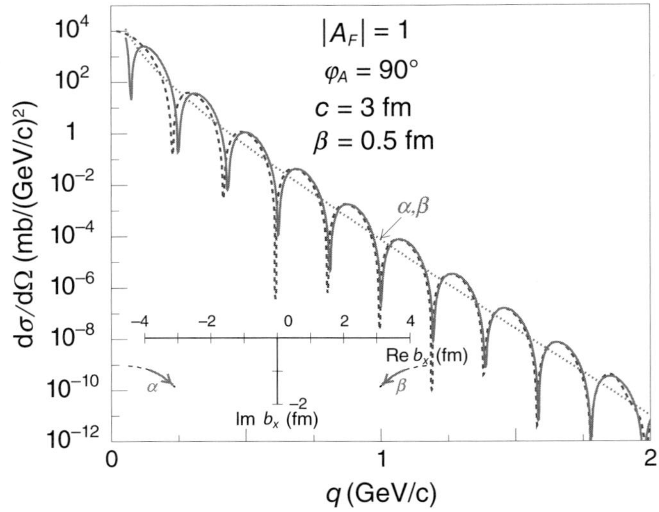

Figure 8.10 Similar to Fig. 8.9, for $\varphi_A = 90°$.

closely for modest values of q and then hardly move at all as q increases. The effect of the poles of $X_{\rm F}$ then is to stabilize the slope and the oscillation period of the corresponding elastic differential cross section.

Once again, characteristically, the cross section oscillations for the purely imaginary interaction, Fig. 8.10, are undamped, while those for the complex interaction, Fig. 8.9, are damped. From these and our earlier results it is clear that the presence of undamped oscillations in a logarithmic plot of a differential cross section indicates that the interaction is greatly dominated by the imaginary or absorptive part. Another characteristic of purely absorptive interactions is that within the context of diffraction theory the differential cross sections have periodic zeros. Corrections to the diffraction theory, Coulomb field effects, and various spin-dependent effects will tend to fill in the zeros, however, and leave minima that are often considerably less pronounced. Figs. 8.9 and 8.10 both indicate that the asymptotic theory is fairly accurate away from the forward direction for the "Fermi" phase shift function. Actually, the result of the numerical evaluation of the diffraction integral is based on the symmetrized phase shift function, Eq. (8.8). Without such a symmetry (or evenness, see Section 8.1), the cross section would start to deviate from this regular pattern around $q = 1.5$ GeV/c.

As is seen from Eq. (8.31), the stationary points depend on the strength of the interaction through the parameter $\lambda = A_{\rm F}/2q\beta$. On the other hand, $X_{\rm F}(b_{jx})$ and $X''_{\rm F}(b_{jx})$ are functions of both $A_{\rm F}$ and λ, and thus the cross section will depend on $A_{\rm F}$ and λ separately. As a result, the period of oscillation of the differential cross section and the slope of its envelope will depend not only on c and β, but on the

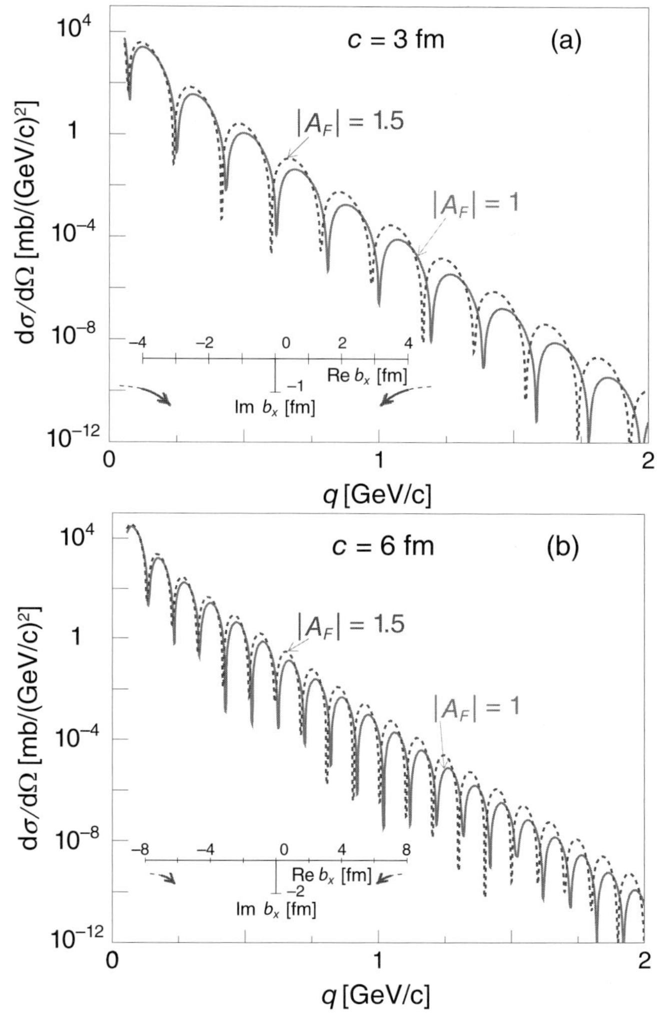

Figure 8.11 Differential cross sections for the "Fermi" phase shift function, $\chi_F(b) = A_F/\{1 + \exp[(b - c)/\beta]\}$. Two radius parameters are considered, (a) $c = 3$ fm, (b) $c = 6$ fm. In both cases, $\beta = 0.5$ fm and A_F is purely imaginary. The figure illustrates the fact that not only the geometry, but also the strength of the interaction is of importance for the shape of the differential cross section.

magnitude of A_F as well. This dependence is illustrated in Fig. 8.11 for interactions of two different radii. Increasing the strength of the interaction is seen to decrease the oscillation period of the cross section. The same dependence is evident in our earlier Fig. 7.7.

We have seen that at small momentum transfers three different phase shift functions, X_{exp}, X_{cosh}, and X_F may lead to the same cross section. Obviously, at sufficiently large momentum transfer, one may distinguish between them. This is

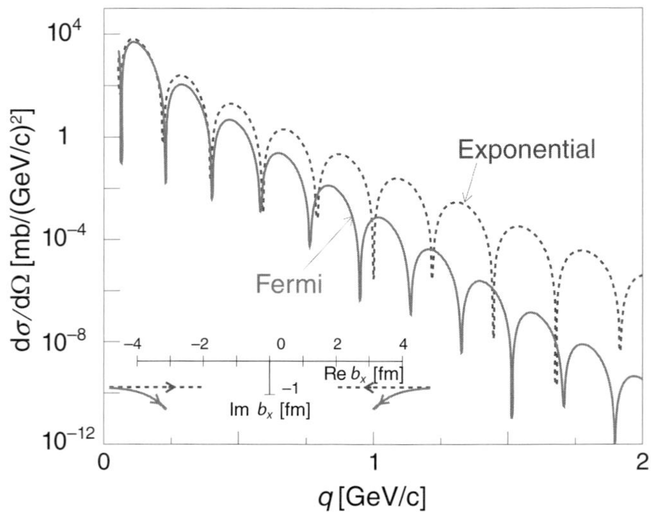

Figure 8.12 A comparison of the differential cross sections resulting from two phase shift functions that coincide at large parameters, but that are different in the interior, $\chi_{\rm exp}(b) = A_{\rm exp}\exp(-b/\beta)$ (dashed), and $\chi_{\rm F}(b) = A_{\rm F}/\{1+\exp[(b-c)/\beta]\}$ (solid). The parameters satisfy Eq. (8.35). Further, $c = 3$ fm, $\beta = 0.5$ fm, and $A_{\rm F} = 2i$. The trajectories of stationary points are given in the insert.

illustrated in Fig. 8.12, where we compare the cross sections corresponding to $X_{\rm exp}$ and $X_{\rm F}$, for parameters satisfying Eq. (8.35), and with $c = 3$ fm, $\beta = 0.5$ fm, and $A_{\rm F} = 2i$. We see that the oscillations in the cross sections corresponding to $X_{\rm F}$ and $X_{\rm exp}$ start to get out of phase and differ in slope when

$$q \gtrsim \frac{|A_{\rm F}|}{2\beta}, \tag{8.41}$$

which corresponds to $q \gtrsim 0.4$ GeV/c. (A similar comparison between $X_{\rm exp}$ and $X_{\rm cosh}$, with $A_{\rm exp} = 2A_{\rm cosh}$, would lead to an agreement between the two out to $q \leq |A_{\rm cosh}|/2\beta$.)

We conclude this section with a comparison of the potentials corresponding to the phase shift functions studied in Fig. 8.12. These are easily determined numerically from Eq. (2.6a). As shown in Fig. 8.13, for a choice of parameters satisfying Eq. (8.35), they are quite similar at large values of r, but differ greatly for smaller radii.

For applications to proton–nucleus scattering, one often considers an optical potential proportional to a Woods–Saxon density function $\rho_{\rm WS}$. If we assume a proportionality between the optical potential and the density, then the density, $\rho[T_{\rm F}]$ corresponding to a “Fermi” thickness function,[1] $T_{\rm F}(b)$ is readily determined

[1] If the range of interaction can be neglected, this is equivalent to a “Fermi” phase shift function, $\chi_{\rm F}(b_x, b_y)$.

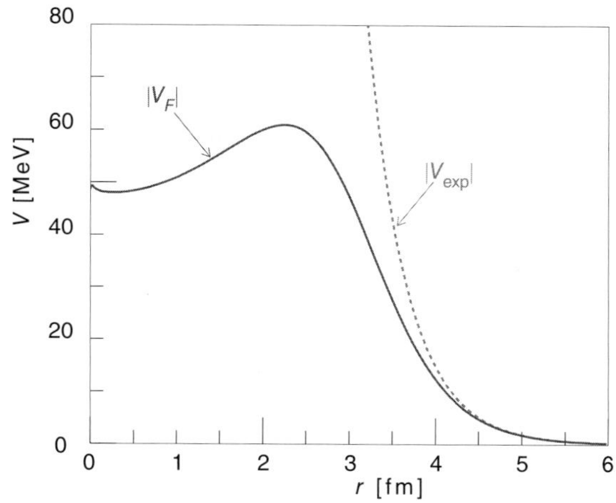

Figure 8.13 Moduli of the potentials corresponding to the exponential and "Fermi" phase shift functions considered in Fig. 8.12.

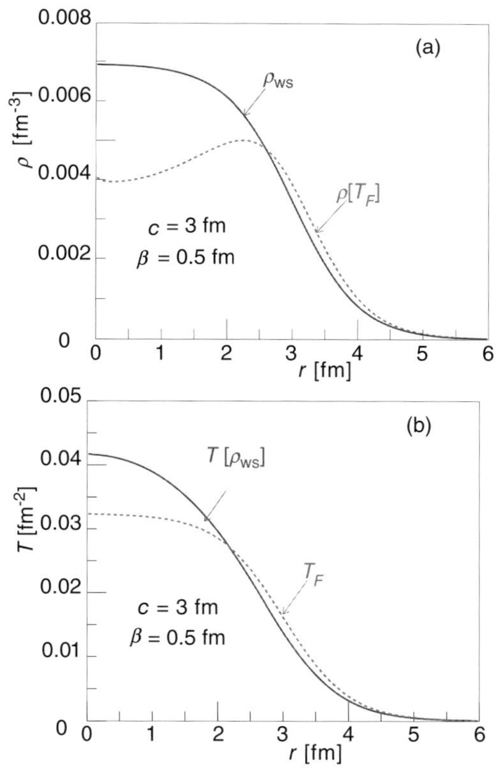

Figure 8.14 (a) A comparison of the Woods–Saxon density, $\rho_{\mathrm{WS}}(r) = \rho_0/\{1 + \exp[(r-c)/\beta]\}$, with the density corresponding to the "Fermi" phase shift function (or a "Fermi" thickness function), $\rho[T_{\mathrm{F}}]$. The normalization is $\int d^3r\rho(r) = 1$. The shape parameters, c and β, are the same in both cases. (b) A comparison of the thickness function corresponding to the Woods–Saxon density, $T[\rho_{\mathrm{WS}}]$, with the "Fermi" thickness function, $T_{\mathrm{F}} = T_0/\{1 + \exp[(b-c)/\beta]\}$. The normalization is $\int d^2bT(b) = 1$. Again the shape parameters are the same.

from the Abel integral equation, Eq. (2.6c). These densities (both normalized to $\int d^3r\rho(r) = 1$) are compared in Fig. 8.14a for $c = 3$ fm and $\beta = 0.5$ fm. For $r \gtrsim c$ (i.e., in the region that is most significant in determining the cross section) they are rather similar in shape, but differ somewhat in normalization. We also note that $\rho[T_F]$ has an enhancement in the region of the surface. Since such an enhancement is sometimes encountered in shell-model calculations [4,18] of nuclear densities, there could be instances in which $\rho[T_F]$ better represents the nuclear density than ρ_{WS}.

For completeness, we also compare in Fig. 8.14b the corresponding thickness functions, $T[\rho_{WS}]$ and T_F, which are quantities more closely related to the cross section. The fact that these are more similar than the densities from which they are derived indicates one reason why nuclear cross sections depend only weakly on the behavior of the density function at small radii.

9

Sharper Edges

In this chapter we briefly comment on phase shift functions having sharper edges than those considered in Chapters 7 and 8. As we shall see, some qualitatively new features appear.

9.1 Quartic Edge

Consider the phase shift function

$$\chi_4(b_x, b_y) = A_4 \exp(-b^4/\beta^4), \tag{9.1}$$

with

$$X_4(b_x) = A_4 \exp(-b_x^4/\beta^4). \tag{9.2}$$

The stationary points will be given by

$$X_4'(b_x) = -4A_4(b_x^3/\beta^4)\exp(-b_x^4/\beta^4) = q. \tag{9.3}$$

Let A_4 be purely imaginary. The loci of stationary points, which satisfy $\mathrm{Im}\, X_4'(b_x) = 0$, $\mathrm{Re}\, X_4'(b_x) > 0$, are given in Fig. 9.1. It is instructive to compare with the Gaussian phase shift function of Section 7.1. As was the case with the Gaussian function, when $q \to 0$, the solutions satisfy

$$\mathrm{Re}\, b_{jx} \to \pm\infty, \quad \mathrm{Im}\, b_{jx} \to 0, \quad q \to 0, \tag{9.4}$$

i.e., they come in from infinity, along the real axes. However, as $q \to \infty$, the points of stationary phase move out along the asymptotes

$$b_{\alpha x} \sim |b_{\alpha x}| \exp(-3i\pi/4), \quad b_{\beta x} \sim |b_{\beta x}| \exp(-i\pi/4), \quad q \to \infty. \tag{9.5}$$

The reason for this movement is that this is where the argument of the exponential attains a large and positive real part.

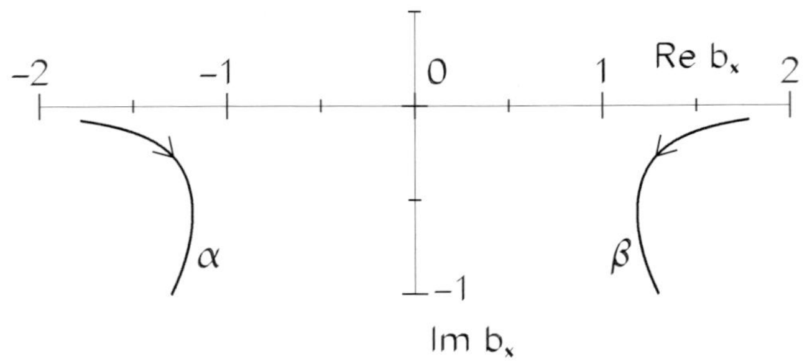

Figure 9.1 Trajectories $b_{\alpha x}$ and $b_{\beta x}$ of stationary impact vectors $\boldsymbol{b}_j = (b_{jx}, 0)$ for the phase shift function Eq. (9.2). The arrows indicate the sense in which the stationary points move as the momentum transfer increases.

This behavior is thus quite different from that observed in Fig. 7.5 for the Gaussian function. As the momentum transfer q increases, starting from some small value, $|\operatorname{Re} b_{\alpha x}|$ and $|\operatorname{Re} b_{\beta x}|$ will first decrease, reach a minimum, and then start to grow again. As a consequence, *the corresponding differential cross section will have oscillations with a period that reaches a maximum, and then becomes shorter again.*

However, the change takes place quite slowly with increasing momentum transfer. From Eq. (9.3), it follows that

$$\frac{db_x}{dq} \underset{q\to\infty}{\sim} \frac{\beta^4}{4q|b_x|^3} e^{-i\pi/4}, \tag{9.6}$$

or

$$b_x \underset{q\to\infty}{\sim} \beta[\log(q\beta)]^{1/4} e^{-i\pi/4}. \tag{9.7}$$

The asymptotic period shortening is so gradual that the "outer" points of the trajectories given in Fig. 9.1 lie at 10 TeV/c.

Another aspect of this sharper edge is the fact that for increasing momentum transfer, the stationary points also move away from the real axis. In the absorptive case, for purely, imaginary A_4, this leads to a steeper envelope of the oscillatory pattern. This is illustrated in Fig. 9.2, where we show the differential cross section for $A_4 = 5i$, $\beta = 1$ fm and a very large range of momentum transfers, namely out to $q = 10$ GeV/c and superimpose the corresponding cross section from $q = 20$ GeV/c to $q = 30$ GeV/c (in steeper curve), multiplied by a scaling factor 10^{50}, in order to better compare the slopes and period of oscillation. The steepening of the envelope is quite pronounced, but in this range of momentum transfers, there is no visible change of the period of oscillation.

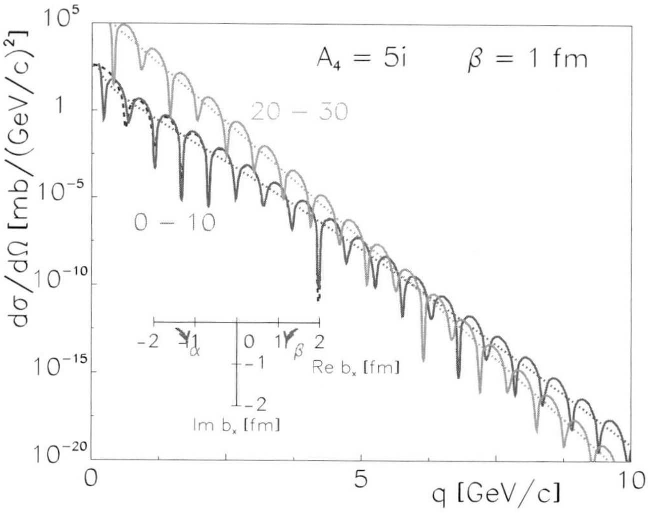

Figure 9.2 Differential cross section for the phase shift function Eq. (9.2). The inserts show the trajectories of stationary points for 0.15 GeV/c $\leq q \leq$ 30 GeV/c (dotted). The parts that correspond to the cross section plots, 0.2 GeV/c $\leq q \leq$ 10 GeV/c are shown as solid curves.

For $q < 7$ GeV/c, we also display the differential cross section, as determined from a numerical integration of the diffraction integral, Eq. (2.2). The agreement is excellent, apart from angles very close to the forward direction.

9.2 Sharper Edges

As an example of an even sharper edge, we consider

$$\chi_6(b_x, b_y) = A_6 \exp(-b^6/\beta^6), \tag{9.8}$$

with

$$X_6(b_x) = A_6 \exp(-b_x^6/\beta^6). \tag{9.9}$$

The stationary points given by

$$X_6'(b_x) = -6A_6(b_x^5/\beta^6)\exp(-b_x^6/\beta^6) = q, \tag{9.10}$$

are shown in Fig. 9.3 for purely imaginary A_6. By an obvious generalization of Eqs. (9.6) and (9.7), we find that the asymptotic period shortening is extremely slow,

$$b_x \underset{q\to\infty}{\sim} \beta[\log(q\beta)]^{1/6}\, e^{-i\pi/6}. \tag{9.11}$$

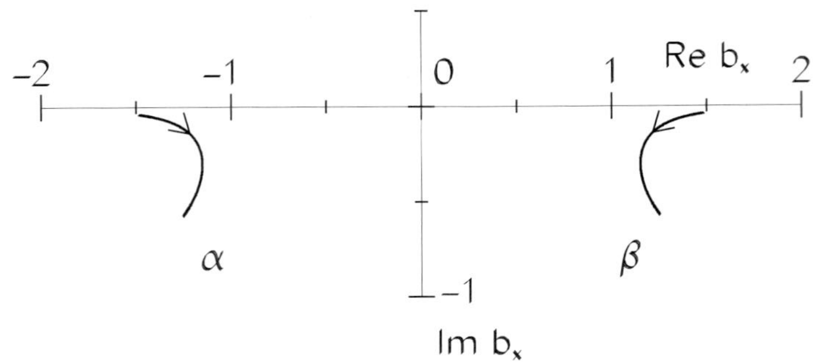

Figure 9.3 Trajectories $b_{\alpha x}$ and $b_{\beta x}$ of stationary impact vectors $\boldsymbol{b}_j = (b_{jx}, 0)$ for the phase shift function Eq. (9.9). The arrows indicate the sense in which the stationary points move as the momentum transfer increases.

In conclusion, we note that these sharper edges can lead to qualitatively different features of the oscillatory pattern of the differential cross section. In particular, the period may reach a maximum, and then decrease again. However, in order to see such phenomena, very large momentum transfers might be required.

10

Phase Shift Functions with Other Analytic Properties

The family of singular phase shift functions studied in Chapter 8 all have a simple pole associated with the exponential edge, i.e., they have pole singularities at $b_x = \pm c - i\pi\beta$, whereas those considered in Chapters 7 and 9 have essential singularities. A focus on these singularities is instructive, since it conveys information on how the points of stationary phase move at large momentum transfers.

For example, with the "Fermi" phase shift function

$$X(b_x) = \frac{A}{1 + \exp[(b - c)/\beta]}, \tag{10.1}$$

we will asymptotically (when $\exp[(b - c)/\beta] \to -1$) find

$$X(b_x) \sim \frac{-A\beta}{b_x - b_x^{\text{pole}}}, \quad b_x \to b_x^{\text{pole}}, \tag{10.2}$$

with

$$q = X'(b_x) \sim \frac{A\beta}{(b_x - b_x^{\text{pole}})^2}. \tag{10.3}$$

Thus, the derivatives will be more singular; they have a *double* pole. The points of the stationary phase will, as $q \to \infty$, asymptotically approach the pole singularity as the inverse square root of q:

$$b_x = b_x^{\text{pole}} \pm \sqrt{\frac{A\beta}{q}}. \tag{10.4}$$

For the case of a purely absorptive phase shift function, $A = i|A|$, the approach to the singularity is seen to be at an angle of $\pm\pi/4$, as we have seen in several examples in Chapter 8. This angle of $\pm\pi/4$ means that the period and exponential slope (determined by real and imaginary parts of b_x, respectively) will change with increasing momentum transfer at the same relative rate.

This behavior is however not unique, as we will discuss in this chapter. The asymptotic behavior of the exponential slope of the differential cross section and the period of oscillation both depend on how the stationary point approaches the singularity. But they need not approach the singularity as $1/\sqrt{q}$, and they need not even approach it at the same rate.

10.1 Other Powers of the "Fermi" Phase Shift Functions

A simple modification of the "Fermi" phase shift function, in which we raise the denominator to some real power $p - 1$,

$$X_p(b_x) = \frac{A}{\{1 + \exp[(b - c)/\beta]\}^{p-1}}, \qquad \mathrm{Re}\, b_x > 0,$$
$$X_p(-b_x) = X_p(b_x), \tag{10.5}$$

will make the stationary point approach the singularity as

$$X'_p(b_x) \sim \frac{1}{(b_x - b_x^{\mathrm{sing}})^p}, \tag{10.6}$$

with $b_x^{\mathrm{sing}} = \pm c - i\pi$.

10.2 "Derivative-Based" Phase Shift Functions

The discussion is actually simpler in terms of the derivative of the phase shift function, rather than in terms of the phase shift function itself. However, in order to make sure the phase shift function has a reasonably well-defined extension, we will make use of the cut-off provided by the "Fermi" function. On the real axis, we shall study a generalization of the "Fermi" phase shift function:

$$X'(b_x) = -\frac{(A/\beta)(b_x/c)^{1+2\nu}}{(1 + \exp[(b_x - c)/\beta])^p}, \qquad b_x > 0,$$
$$X'(-b_x) = X'(b_x). \tag{10.7}$$

Thus, we let the *derivative* have a singularity of order p, where p need not even be an integer. We allow the power ν to be zero or a positive integer in order that the phase shift function $X(b_x)$ be even. Asymptotically, for $q \to \infty$, the singularity of $X'(b_x)$ will be probed, where

$$X'(b_x) \sim -\frac{(A/\beta)\beta^p(1 - i\pi\beta/c)^{1+2\nu}}{(b_x^{\mathrm{sing}} - b_x)^p}, \qquad \mathrm{Re}\, b_x > 0, \tag{10.8}$$

with $b_x^{\text{sing}} = c - i\pi\beta$, and similarly the singularity at $\operatorname{Re} b_x < 0$. It is convenient to introduce some notation for the last factor in the numerator of Eq. (10.8), stemming from the factor b_x/c in the numerator of $X'(b_x)$:

$$1 - i\pi\frac{\beta}{c} = \sqrt{1 + \left(\pi\frac{\beta}{c}\right)^2}\, e^{-i\eta} \equiv \xi e^{-i\eta}. \tag{10.9}$$

Equating now the expression Eq. (10.8) to the momentum transfer q, we find

$$b_x \simeq b_x^{\text{sing}} - \left(\frac{-A\xi e^{-i\eta}}{q\beta}\right)^{1/p} \beta, \tag{10.10}$$

where we have taken $\nu = 0$, but it is clear how this is modified for $\nu > 0$. It is also immediately clear that the power p controls how fast the point of stationary phase approaches the singularity as $q \to \infty$.

For simplicity of the discussion, we shall in this chapter only consider purely real or purely absorptive phase shift functions. Then, we note that a non-zero power ν can to some extent modify the phase of $b_x - b_x^{\text{sing}}$ in the complex plane. This, in turn, determines the relative rates of change of the period of oscillation versus the slope of the exponential envelope.

For $p = 1$, the approach is determined by $1/q$, and in the absence of the phase η, it would for an absorptive phase shift function ($A = i|A|$) be purely imaginary, i.e., only the exponential slope would change, not the period of oscillation:

$$b_x \simeq b_x^{\text{sing}} + \frac{i|A|}{q}. \tag{10.11}$$

For a more accurate analysis, keeping the last factor of Eq. (10.8), this approach would be rotated by the angle $-\eta$, which for $c = 3$ fm and $\beta = 0.5$ fm would amount to -27.6°, producing also some lengthening of the period.

For $p = 3$, on the other hand,

$$b_x = b_x^{\text{sing}} - \left(\frac{-A\xi e^{-i\eta}}{q\beta}\right)^{1/3} \beta, \tag{10.12}$$

which in the absorptive case becomes

$$b_x = b_x^{\text{sing}} + \exp\left[\frac{i}{3}\left(-\frac{\pi}{2} + 3\pi + 2n\pi - \eta\right)\right]\left(\frac{|A|\xi}{q\beta}\right)^{1/3} \beta. \tag{10.13}$$

As announced, the point of stationary phase approaches the asymptotic value much more slowly. For $\eta \to 0$, it would approach at an angle of $\pm 5\pi/6$ (for $n = 0$ in the above expression). For $n \neq 1$, there are additional solutions for the phase of $b_x - b_x^{\text{sing}}$, but those solutions b_x are not along the path of integration.

This chapter will illustrate some phenomena related to these other singularities.

10.2.1 Soft-Exponential Phase Shift Function

We shall first discuss a phase shift function which is related to those phase shift functions with an exponential edge, but in analytical terms has a somewhat softer edge. As outlined above, we will define this in terms of its derivative. Let

$$f_1(b_x) \equiv \frac{-(A/\beta)(b_x/c)}{1+\exp[(b_x-c)/\beta]}, \quad b_x > 0, \tag{10.14}$$

and take

$$\begin{aligned} X'(b_x) &= f_1(b_x), \qquad b_x > 0, \\ X'(b_x) &= -f_1(b_x), \quad b_x < 0. \end{aligned} \tag{10.15}$$

In order to calculate the cross section, we also need the phase shift function itself. This can be obtained by integration:

$$\begin{aligned} X(b_x) &= -\int_{b_x}^{\infty} \mathrm{d}t\, X'(t) = -\int_{b_x}^{\infty} \mathrm{d}t\, f_1(t), \quad \mathrm{Re}\, b_x > 0, \\ X(-b_x) &= X(b_x). \end{aligned} \tag{10.16}$$

For b_x off the real axis, this will be an integral along a complex path, it can easily be expressed as a sum of two integrals, one along a path parallel to the imaginary axis, and one along the real axis. Another example will be based on

$$f_3(b_x) \equiv \frac{-(A/\beta)(b_x/c)^3}{1+\exp[(b_x-c)/\beta]}, \quad b_x > 0, \tag{10.17}$$

where $X'(b_x)$ and $X(b_x)$ are obtained by the same equations Eqs. (10.15 and 10.16), but with $f_3(b_x)$ substituted for $f_1(b_x)$. This gives a more flat phase shift function in the interior, for $|b_x| < c$, but more of a tail for $|b_x| > c$.

While the derivative of the "Fermi" phase shift function, given by Eq. (8.7), has a double pole at $b_x = \pm c - i\pi\beta$, the present one has only a single pole. As a result, with increasing momentum transfers, the stationary points will approach these singularities at a faster rate. The corresponding lengthening of the period of oscillation will set in at a lower value of momentum transfer, and will be faster. This is seen from the asymptotic relation

$$q = X'(b_x) \simeq \frac{A\xi e^{-i\eta}}{b_x - b_x^{\mathrm{pole}}}, \tag{10.18}$$

with $b_x^{\mathrm{pole}} = \pm c - i\pi\beta$. Inverting, one finds

$$b_x = b_x^{\mathrm{pole}} + \frac{A\xi e^{-i\eta}}{q}. \tag{10.19}$$

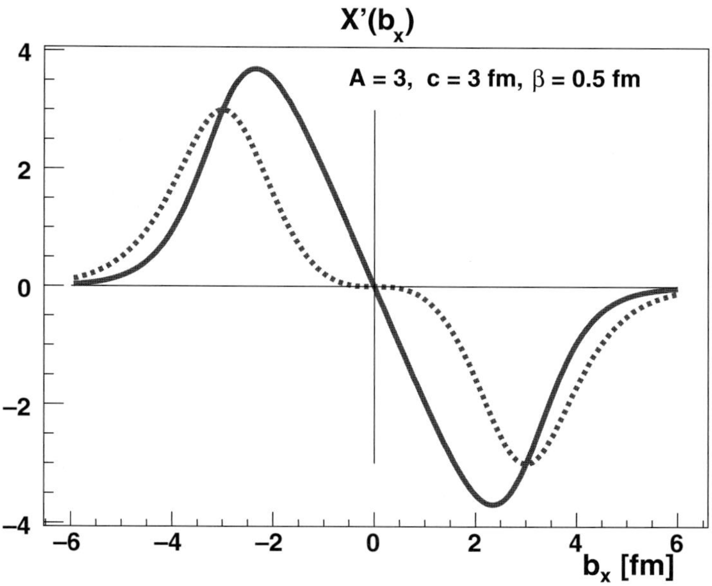

Figure 10.1 Derivatives of a phase shift function having a softer edge than the "Fermi" one. Solid curve: based on f_1, Eq. (10.15); dashed curve: based on f_3, Eq. (10.17).

For comparison, we recall that for the "Fermi" phase shift function discussed in Section 8.4, the approach to the pole singularity is like $1/\sqrt{q}$; see Eqs. (8.39) and (8.40).

For shape parameters $c = 3$ fm and $\beta = 0.5$ fm, together with a strength $A = 1$, we show in Figs. 10.1 and 10.2 this derivative $X'(b_x)$ and the phase shift function $X(b_x)$. While the region around $b_x = c$ (the edge) of the phase shift function is similar to the "Fermi" phase shift function, the interior is more peaked around $b_x = 0$.

Possible locations of points of stationary phase are given by the lines in the complex b_x-plane, where $X'(b_x)$ is real and positive. Such lines are identified in the two "maps" in Fig. 10.3. Here we display regions where $\mathrm{Re}\, X'(b_x) > 0$ (white) together with the complementary region where $\mathrm{Re}\, X'(b_x) < 0$ (shaded), as well as where $\mathrm{Im}\, X'(b_x) = 0$ (dashed lines). Both of these maps are based on derivatives $X'(b_x)$ having simple poles at $b_x = \pm c - i\pi\beta$ (indicated as "bullets", for $c = 3$ fm, $\beta = 0.5$ fm), one vanishes linearly at the origin, $X'(b_x) \sim b_x$ as $b_x \to 0$, the other in a cubic fashion, $X'(b_x) \sim b_x^3$ as $b_x \to 0$. The importance of these powers lies in the phase of the factor $(\xi e^{-i\eta})^{1+2\nu}$ present in the asymptotic expression Eq. (10.18). With the cubic factor of $f_3(b_x)$, we see that the stationary points approach the singularity practically parallel to the real axis. This means that

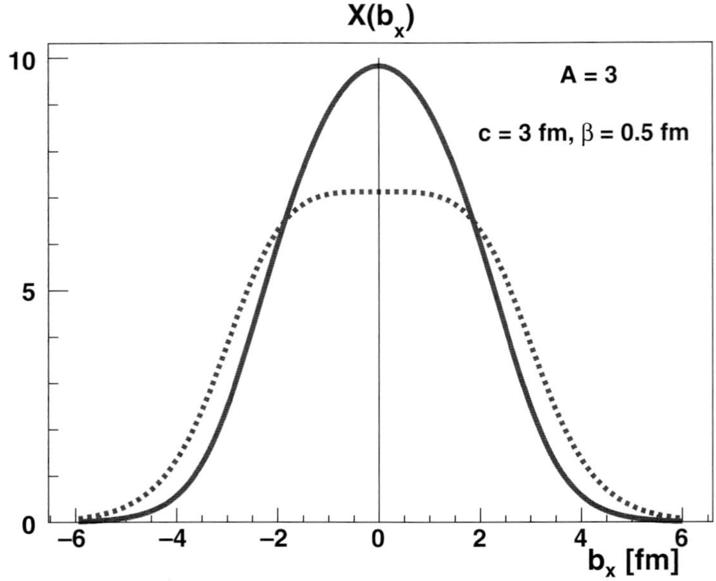

Figure 10.2 Phase shift functions $X(b_x)$ corresponding to the derivatives shown in Fig. 10.1.

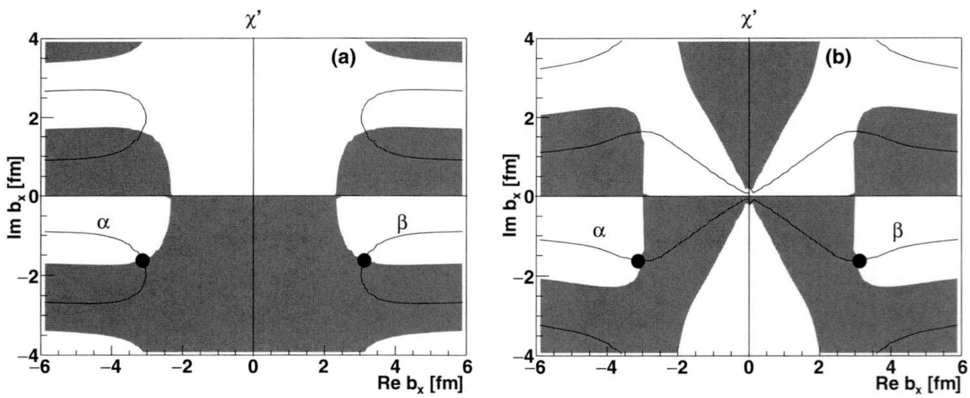

Figure 10.3 Regions of positive (white) and negative (shaded) values of $\operatorname{Re} X'(b_x)$, shown in the complex b_x-plane for the purely absorptive case. The figures are analogous to Fig. 7.2. Also indicated as dashed lines, are the locations where $\operatorname{Im} X'(b_x) = 0$. (a) $X'(b_x)$ determined by Eq. (10.14). (b) $X'(b_x)$ determined by Eq. (10.17).

asymptotically, the main change will be in the period of the oscillation, not so much in the exponential envelope.

For the purely absorptive case, with $A = 3i$, we obtain the differential cross section shown in Fig. 10.4. We here compare the differential cross section based

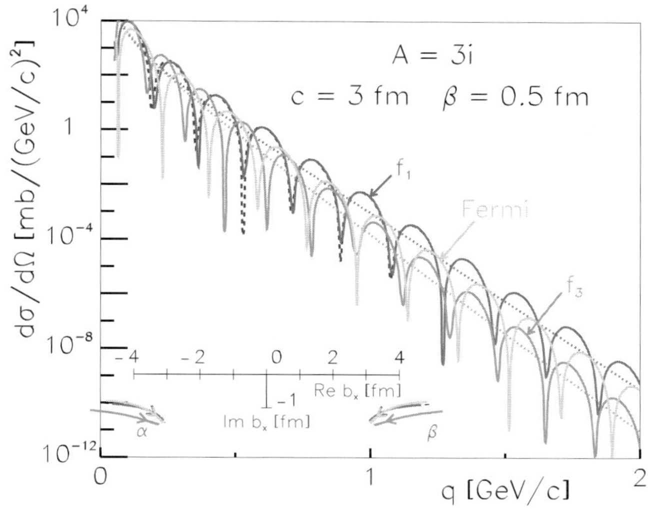

Figure 10.4 Differential cross section corresponding to the phase shift function whose derivative is given by Eq. (10.15). The different solid curves refer to f_1, f_3 and the "Fermi" phase shift function. Dashed (small q): numerical integration for the case of f_1. The inset shows the relevant trajectories of stationary points.

on the function f_1 (with $\nu = 0$) with that based on the function f_3 (with $\nu = 1$), for the same shape (c, β) and strength parameters. For a quantitative comparison of these differential cross sections, we can approximate the exponential envelope and the oscillations of the differential cross section as

$$\frac{d\sigma}{d\Omega} \sim \exp\left(-\frac{q}{Q}\right) \times \cos^2\left(\frac{\pi q}{P} + \phi\right)^2 \tag{10.20}$$

and find the parameters Q (slope) and P (period) in different ranges of momentum transfer q. The results of such fits are quoted in Table 10.1. It should be noted that these quantities depend not only on the shape parameters c and β and the functional form, but also on the strength of the interactions, here taken as $|A| = 3$.

Because of the additional phase η, these numbers do not quite agree with the qualitative discussion above. But the rate of change (with q) is seen to be much faster than for the "Fermi" phase shift function.

10.2.2 Sharp-Exponential Phase Shift Function

As an example of a phase shift function with a higher-order pole near the surface, let its derivative be defined in terms of

$$g_1(b_x) \equiv \frac{-(A/\beta)(b_x/c)}{(1 + \exp[(b_x - c)/\beta])^3}, \quad b_x > 0, \tag{10.21}$$

Table 10.1 *Exponential slopes Q and periods P, for phase shift functions derived from their derivatives, f_1 and f_3, and, for reference, the "Fermi" phase shift function. Two ranges of momentum transfer are considered (with the increment indicated in parenthesis).*

	0.1 GeV/c $< q <$ 0.5 GeV/c		1.5 GeV/c $< q <$ 1.9 GeV/c	
Model	Q	P	Q	P
"Fermi"	0.128	0.172	0.150 (+17%)	0.195 (+13%)
$\nu = 0$	0.129	0.161	0.168 (+30%)	0.193 (+20%)
$\nu = 1$	0.123	0.145	0.145 (+18%)	0.181 (+25%)

which is the same as the function $f_1(b_x)$ of Eq. (10.14), except for the higher power in the denominator. Note that this is *more* singular than the derivative of the "Fermi" phase shift function, that also here will serve as a reference. Again, we take

$$X'(b_x) = g_1(b_x), \qquad b_x > 0,$$
$$X'(-b_x) = -X'(b_x), \tag{10.22}$$

and determine the phase shift function itself according to Eq. (10.16). We shall also compare with phase shift functions based on

$$g_3(b_x) \equiv \frac{-(A/\beta)(b_x/c)^3}{(1+\exp[(b_x-c)/\beta])^3}, \qquad b_x > 0, \tag{10.23}$$

which is analogous to Eq. (10.17).

In Figs. 10.5 and 10.6 we show the derivative $X'(b_x)$ and $X(b_x)$, as defined by Eqs. (10.22) and (10.21) (solid) as well as by Eq. (10.23) (dashed). While the edges are similar, the interior is quite different.

Asymptotically, the points of stationary phase will be determined by

$$b_x \sim b_x^{\text{sing}} + \exp\left[\frac{i}{3}\left(-\frac{\pi}{2}+3\pi+2\pi n-\eta\right)\right]\left(\frac{|A|\xi}{q\beta}\right)^{1/3}, \tag{10.24}$$

$$b_x \sim b_x^{\text{sing}} + \exp\left[\frac{i}{3}\left(-\frac{\pi}{2}+3\pi+2\pi n-3\eta\right)\right]\left(\frac{|A|\xi^3}{q\beta}\right)^{1/3}, \tag{10.25}$$

for g_1 and g_3, respectively.

In Fig. 10.7 we show the corresponding maps of possible trajectories of stationary points, analogous to Fig. 10.3 for the phase shift functions discussed in Section 10.2.1. Two cases are considered, $\nu = 0$ (Fig. 10.7(a) $X'(b_x) \sim b_x/\beta$ as $b_x \to 0$) and $\nu = 1$ (Fig. 10.7(b) $X'(b_x) \sim (b_x/\beta)^3$ as $b_x \to 0$). Each singularity here is surrounded by six lines (dashed) of vanishing Im $X'(b_x)$. Half of these are

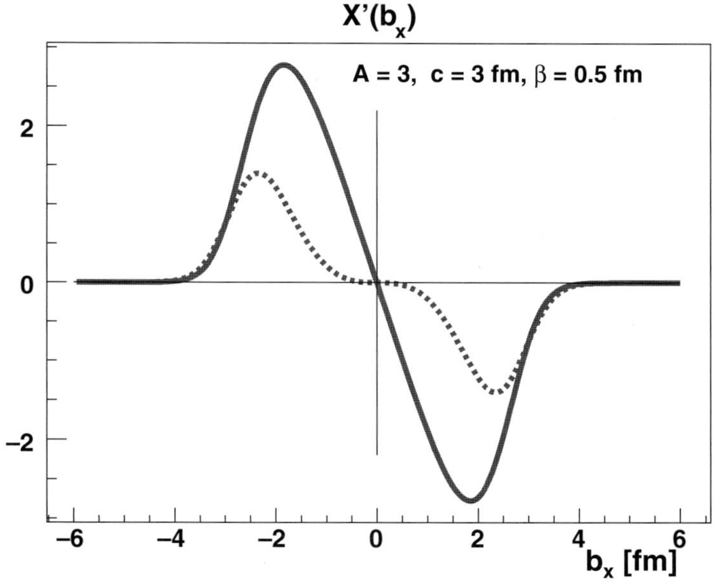

Figure 10.5 The derivative, Eq. (10.22), of phase shift functions based on $g_1(b_x)$ and $g_3(b_x)$. They are more singular than the "Fermi" one.

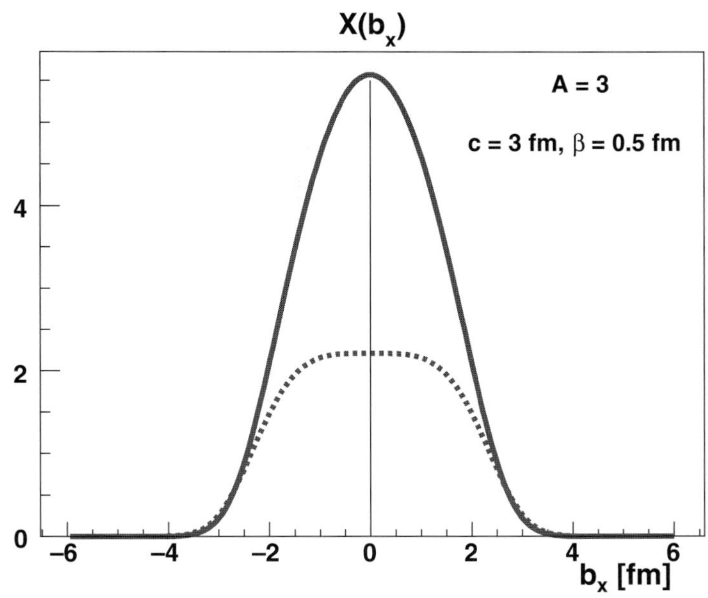

Figure 10.6 The phase shift functions $X(b_x)$ corresponding to the derivatives shown in Fig. 10.5.

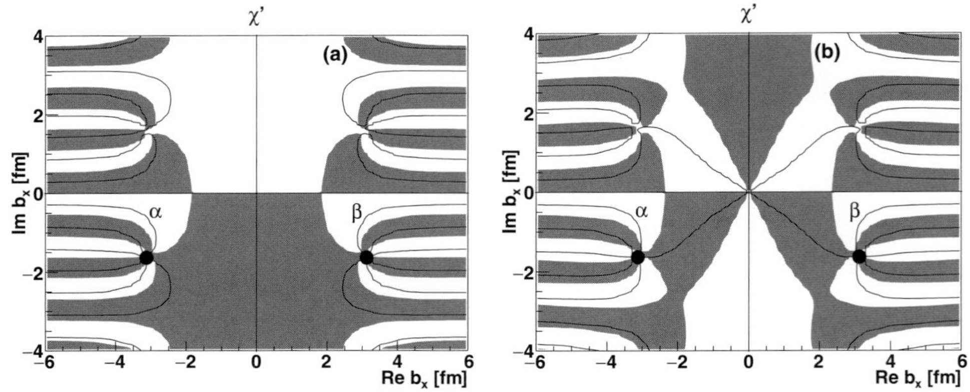

Figure 10.7 Regions of positive (white) and negative (shaded) values of $\mathrm{Re}\, X'(b_x)$, shown in the complex b_x-plane for the purely absorptive case. The figures are analogous to Fig. 10.3. Also indicated as dashed lines, are the locations where $\mathrm{Im}\, X'(b_x) = 0$. (a) $X'(b_x)$ determined by Eq. (10.21). (b) $X'(b_x)$ determined by Eq. (10.23).

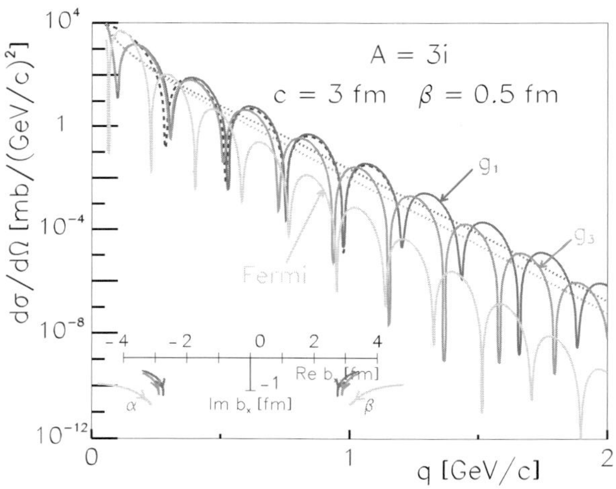

Figure 10.8 Differential cross section (denoted "g_1" and "g_3") corresponding to the sharp "Fermi" phase shift function whose derivatives are given by Eqs. (10.21) and (10.23).

in regions where $\mathrm{Re}\, X'(b_x) > 0$, but only one is actually on the path of integration. That is the one closest to the real axis.

Furthermore, in Fig. 10.8, we compare the differential cross sections of the present model, for both $g_1(b_x)$ and $g_3(b_x)$ with that obtained from the "Fermi" phase shift function, also shown in Figs. 8.10 and 8.12. The faster lengthening of the period is clearly seen in this figure. The inset shows the corresponding trajectories

Table 10.2 *Exponential slopes Q and periods P, for phase shift functions derived from their derivatives, g_1 and g_3, and, for reference, the "Fermi" phase shift function. Two ranges of momentum transfer are considered (with the increment indicated in parenthesis).*

	$0.1\ \mathrm{GeV/c} < q < 0.5\ \mathrm{GeV/c}$		$1.5\ \mathrm{GeV/c} < q < 1.9\ \mathrm{GeV/c}$	
Model	Q	P	Q	P
"Fermi"	0.128	0.172	0.150 (+17%)	0.195 (+13%)
$\nu = 0$	0.194	0.220	0.214 (+10%)	0.224 (+2%)
$\nu = 1$	0.168	0.218	0.180 (+7%)	0.212 (−3%)

of stationary points. They exhibit indeed a faster approach to the pole singularities, as compared with the "Fermi" phase shift function.

The fact that the differential cross section for the "Fermi" phase shift function is more steep, is reflected in Q being smaller for that case. On the other hand, the sharp "Fermi" phase shift function gives longer periods, reflected in a larger value of P. The faster approach to the pole singularities is reflected in P being larger than for the "Fermi" distribution already at low values of q, see Table 10.2.

10.2.3 Other Powers

As a generalization of the shapes defined by Eqs. (10.14) and (10.21), let us consider

$$p_1(b_x) \equiv \frac{-(A/\beta)(b_x/c)}{(1+\exp[(b_x-c)/\beta])^p}, \qquad b_x > 0, \tag{10.26}$$

for p not necessarily an integer. We take the derivative of the phase shift function to be

$$\begin{aligned} X'(b_x) &= p_1(b_x), \qquad b_x > 0, \\ X'(-b_x) &= -X'(b_x), \end{aligned} \tag{10.27}$$

and determine the phase shift function itself according to Eq. (10.16). In analogy with Eqs. (10.17) and (10.23), we shall also compare with phase shift functions with a higher power in the numerator,

$$p_3(b_x) \equiv \frac{-(A/\beta)(b_x/c)^3}{(1+\exp[(b_x-c)/\beta])^p}, \qquad b_x > 0. \tag{10.28}$$

For $p = 1$, we recover the "soft exponential" phase shift functions discussed in Section 10.2.1, whereas for $p = 3$ we recover the "sharp exponential" phase shift functions discussed in Section 10.2.2. For $p = 2$ we obtain a phase shift function whose derivative has the same singularities as the "Fermi" distribution. However, the derivative of the "Fermi" distribution will differ by an additional exponential factor, as well as a different interpretation of the normalization factor χ_0.

For Eq. (10.8), with $X'(b_x) = q$ and for purely absorptive interactions, the approach to the singularity in the fourth quadrant for arbitrary powers p and ν is given by

$$b_x = b_x^{\text{sing}} + \beta \exp\left\{\frac{i}{p}\left[-\frac{\pi}{2} + p\pi + 2\pi n - \eta(1+2\nu)\right]\right\}\left(\frac{|A|}{q\beta}\right)^{1/p} \xi^{(1+2\nu)/p},$$
$$n = 0, 1, 2, \ldots, p-1. \tag{10.29}$$

For the singularities in the fourth quadrant, this can be written schematically as

$$b_x = b_x^{\text{sing}} + \beta\left(\frac{q_0}{q}\right)^{1/p} e^{i\varphi}. \tag{10.30}$$

The envelope of the differential cross section will then depend on q as

$$Q = -\operatorname{Im} b_x^{\text{sing}} - \beta\left(\frac{q_0}{q}\right)^{1/p} \sin\varphi, \tag{10.31}$$

whereas the period will vary as

$$P = \operatorname{Re} b_x^{\text{sing}} + \beta\left(\frac{q_0}{q}\right)^{1/p} \cos\varphi, \tag{10.32}$$

Some powers and the corresponding phases φ are collected in Table 10.3.

10.3 Additional Phases

It is now clear that a variety of asymptotic approaches to the singularities are possible. Another set of approaches is provided by multiplying the derivative $X'(b_x)$

Table 10.3 *Approach of the stationary points to the singularity in the fourth quadrant, for some powers p and phase φ, as defined by Eq. (10.30).*

p	φ
3	$5\pi/6 - \eta(1+2\nu)/3$
2	$-\pi/4 - \eta(1+2\nu)/2$
1	$\pi/2 - \eta(1+2\nu)$
1/2	$\pi - 2\eta(1+2\nu)$
1/3	$3\pi/2 - 3\eta(1+2\nu)$

of Eq. (10.7) by some function which is real and positive on the real axis, but which acquires a different phase in the neighborhood of the singularities. An example of such a factor is $\exp[(b_x - c)/\beta]$. This is obviously real and positive on the real axis, but approaches -1 in the neighborhood of the singularity. Mathematically, this can be described by replacing q by $-q$ in Eq. (10.10). This amounts to changing φ of Table 10.3 by π/p, and is in fact how one recovers the approach $\varphi = \pi/4$ for the "Fermi" phase shift function in the case $p = 2$.

Part III

Composite Phase Shift Functions

11

Coulomb Scattering from Extended Charges

We have seen in the preceding chapters that the theory works accurately for phase shift functions of the sort considered in hadron–nucleus scattering problems. It was also shown in chapter 4 that the asymptotic approach gives an exact result for the Coulomb scattering from a point charge. In the present chapter we shall demonstrate that the method retains its accuracy and presents great practical simplifications in problems involving pure Coulomb scattering by charge distributions. In the next chapter, we will consider combinations of such Coulomb interactions with short-range (nuclear) interactions.

Because of the long range of the Coulomb field, numerical evaluation of the diffraction integral, when charges are present, often requires some special treatment. We shall show, however, that in the asymptotic approach, the infinite range of the Coulomb field needn't pose any problem.

Let us first consider scattering from a distributed charge. For a point-like projectile of charge $Z_1 e$, the Coulomb potential is given in terms of the charge density $\rho_C(\boldsymbol{r})$ as

$$V_C(\boldsymbol{r}) = \frac{Z_1 Z_2 e^2}{4\pi} \int d^3 r' \frac{\rho_C(\boldsymbol{r}')}{|\boldsymbol{r} - \boldsymbol{r}'|}, \tag{11.1}$$

where $\int d^3 r' \rho_C(\boldsymbol{r}') = 1$. The Coulomb phase shift function is then defined as

$$\chi_C(\boldsymbol{b}) = -\frac{1}{\hbar v} \int_{-\infty}^{\infty} dz\, V_C(\boldsymbol{b}, z). \tag{11.2}$$

The unscreened point charge would actually yield a divergent phase shift function. However, a $\boldsymbol{b}$-independent contribution to the phase shift function by the screening charge, however distant it may be, cancels the divergence, leaving only a finite phase attached to the scattering amplitude. This phase, in practice, depends on atomic parameters, and is usually left unmeasured.

11.1 Screened Charge Distribution—An Example

We consider the screened charge distribution

$$\begin{aligned}\rho_C(\boldsymbol{r}) &= \rho_+(r) - \rho_-(r),\\ &= \frac{3}{4\pi R_+^3}\theta(R_+ - r) - \frac{1}{4\pi R_-^2}\delta(r - R_-),\end{aligned} \tag{11.3}$$

with $R_+ < R_-$, and where the positive and negative charge distributions are separately normalized, $\int \mathrm{d}^3 r\rho_+(r) = \int \mathrm{d}^3 r\rho_-(r) = 1$. Hence, the total charge is zero. With

$$\int \mathrm{d}\Omega \frac{1}{|\boldsymbol{r} - \boldsymbol{r}'|} = \frac{4\pi}{\max(r, r')}, \tag{11.4}$$

the potential Eq. (11.1) becomes

$$\begin{aligned}V_C(\boldsymbol{r}) &= Z_1 Z_2 e^2 \left[\frac{1}{r}\int_0^r \mathrm{d}r'\, r'^2[\rho_+(r') - \rho_-(r')] + \int_r^\infty \mathrm{d}r' r'[\rho_+(r') - \rho_-(r')]\right]\\ &= \frac{Z_1 Z_2 e^2}{4\pi}\begin{cases}\dfrac{3}{2R_+} - \dfrac{r^2}{2R_+^3} - \dfrac{1}{R_-}, & r < R_+ < R_-,\\ \dfrac{1}{r} - \dfrac{1}{R_-}, & R_+ < r < R_-,\\ 0, & R_+ < R_- < r.\end{cases}\end{aligned} \tag{11.5}$$

In Fig. 11.1(a) we show the charge distributions Eq. (11.3) and the resulting potential Eq. (11.5) for $R_+ = 1$, $R_- = 5$ (arbitrary units) and $Z_1 Z_2 e^2 = 4\pi$.

The Coulomb phase shift function can be expressed directly in terms of the potential, by Eq. (11.2), using the integral

$$\int_{z_0}^{z_1} \frac{\mathrm{d}z}{\sqrt{b^2 + z^2}} = \log \frac{z_1 + \sqrt{b^2 + z_1^2}}{z_0 + \sqrt{b^2 + z_0^2}}, \tag{11.6}$$

or, alternatively (see Section 11.2), via the charge thickness function

$$T_C(b) = \int_{-\infty}^{\infty} \rho_C\big(\sqrt{b^2 + z^2}\big)\,\mathrm{d}z, \qquad 2\pi \int_0^\infty T_C(b) b\,\mathrm{d}b = 1. \tag{11.7}$$

For the screened charge distribution considered, this thickness function is given by

$$T_C(b) = T_C^+(b) - T_C^-(b), \tag{11.8}$$

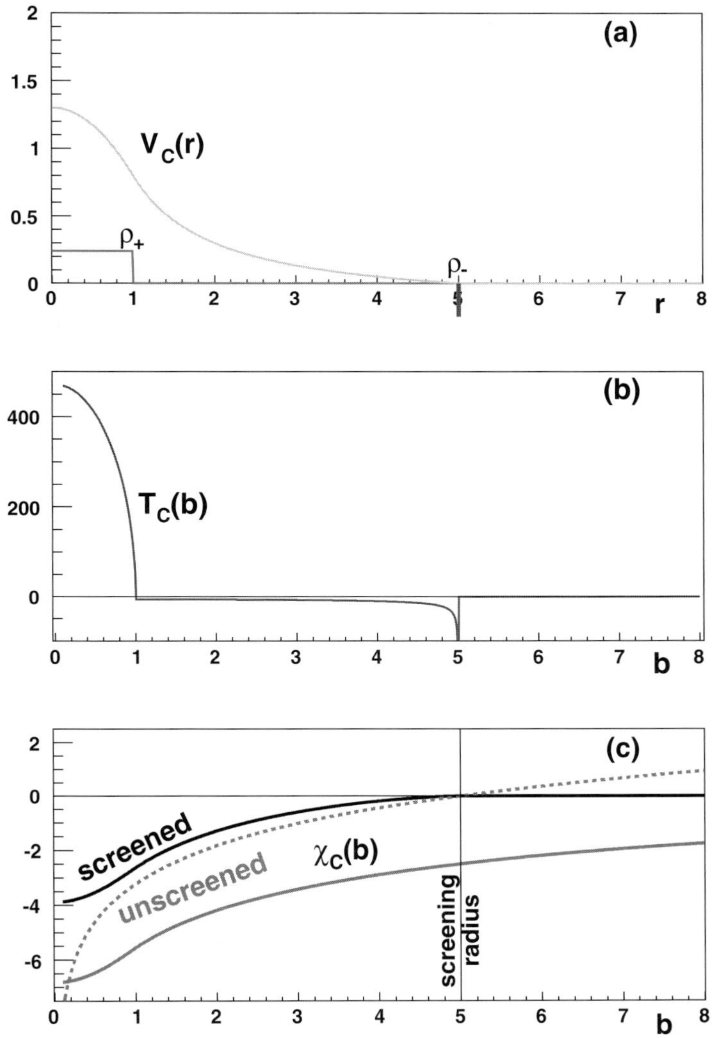

Figure 11.1 Studies of the screened charge distribution defined by Eqs. (11.3), (11.5) and (11.7). (a) Charge distributions ρ_+ and ρ_- (the negative δ-function at $R_- = 5$ is represented by a thick slab) and resulting potential, $V_C(r)$. (b) Thickness function $T_C(b)$. (c) Coulomb phase shift functions: solid: with screening and for finite charge distribution, unscreened (as labeled); dashed: point charge, unscreened.

with

$$T_C^+(b) = \frac{6}{4\pi R_+^2}\sqrt{1-(b/R_+)^2}\,\theta(R_+ - b),$$

$$T_C^-(b) = \frac{2}{4\pi R_-^2}\frac{1}{\sqrt{1-(b/R_-)^2}}\,\theta(R_- - b) \tag{11.9}$$

and shown in Fig. 11.1(b).

In Fig. 11.1(c) we compare the resulting Coulomb phase shift function with two others, both corresponding to unscreened charges. One of them represents a point charge (dashed, and "normalized" to coincide at the screening radius), the other to a distributed charge (solid). Outside of the positive charge ($b > R_+$), the two phase shift functions corresponding to an unscreened charge (point-like or distributed), are seen to only differ by a constant.

11.2 Screened Charge Distribution—Generalities

For a spherically symmetric charge distribution, one gets the familiar result for the phase shift function [11]

$$\chi_{\mathrm{C}}(b_x, b_y) = 4\pi\eta \left[\log b \int_0^b T_{\mathrm{C}}(b')b'\, \mathrm{d}b' + \int_b^\infty T_{\mathrm{C}}(b') \log b' b'\, \mathrm{d}b' \right], \tag{11.10}$$

where

$$b = \sqrt{b_x^2 + b_y^2}, \tag{11.11}$$

and where η is the Sommerfeld parameter,

$$\eta = \frac{Z_1 Z_2 \alpha}{\hbar v} \equiv \frac{Z_1 Z_2 e^2}{4\pi \hbar v}. \tag{11.12}$$

In Eq. (11.10), the logarithms do not refer to the scale of b. In fact, a scale R only contributes a constant amount $-2\eta \log(R)$ (compare Eq. (4.5)), which only modifies the scattering amplitude by an overall constant complex phase.

As seen previously, for cases with azimuthal symmetry, we need only the phase shift function evaluated in the scattering plane, i.e., for $b_y = 0$. Thus,

$$X_{\mathrm{C}}(b_x) \equiv \chi_{\mathrm{C}}(b_x, 0). \tag{11.13}$$

Up to an overall additive constant, as discussed above, one can thus write

$$X_{\mathrm{C}}(b_x) = 4\pi\eta \left[\log b_x \int_0^{b_x} T_{\mathrm{C}}(x)x\, \mathrm{d}x + \int_{b_x}^\infty T_{\mathrm{C}}(x) \log x\, x\, \mathrm{d}x \right], \quad \mathrm{Re}\, b_x > 0, \tag{11.14}$$

$$X_{\mathrm{C}}(-b_x) = X_{\mathrm{C}}(b_x). \tag{11.15}$$

For an arbitrary charge distribution, the stationary points will then be given in terms of $T_{\mathrm{C}}(b_x)$ as roots of either of the equations

$$X_{\mathrm{C}}'(b_x) = 4\pi\eta b_x^{-1} \int_0^{b_x} T_{\mathrm{C}}(x)x\, \mathrm{d}x = q, \quad \mathrm{Re}\, b_x > 0,$$

or

$$X_C{}'(b_x) = 4\pi\eta b_x^{-1}\int_0^{-b_x} T_C(x)x\,dx = q, \qquad \mathrm{Re}\,b_x < 0. \tag{11.16}$$

If we consider next the second derivative,

$$X_C{}''(b_x) = 4\pi\eta b_x^{-1}\left[T_C(b_x)b_x - b_x^{-1}\int_0^{b_x} T_C(x)x\,dx\right], \qquad \mathrm{Re}\,b_x > 0, \tag{11.17}$$

$$X_C{}''(-b_x) = X_C{}''(b_x), \tag{11.18}$$

we note immediately an important difference between the scattering from a point charge and that from a distributed charge. Whereas in the former case there is no rainbow point, there will in the latter case always be at least one rainbow singularity for a monotonically decreasing $T_C(b_x)$. It will occur at a value of the impact parameter $b_{x,R}$ for which the function $T_C(b_x)b_x$ is equal to its average over the interval from 0 to the points $b_{x,R}$,

$$T_C(b_{x,R})b_{x,R} = b_{x,R}^{-1}\int_0^{b_{x,R}} T_C(x)x\,dx. \tag{11.19}$$

This condition yields positive and negative solutions for $b_{x,R}$ which are equal in magnitude. Only one of the solutions, however, is consistent with the condition Eq. (11.16); it is the one for which $X_C{}'$ is positive and $b_{x,R}$ has the same sign as η.

In contrast to the phase shift functions that we have encountered earlier, which were given in terms of some explicit, simple function, the present one is defined in terms of an integral along some non-trivial path in the complex b_x-plane. When the thickness function $T_C(b_x)$ has singularities in the complex b_x-plane, such integrals will depend on the path followed. We will return to this issue in Section 11.5.

11.3 A Simple Charge Distribution

As a particularly elementary example of an extended charge distribution, let us consider the one which generates the phase shift function

$$\chi_1(\boldsymbol{b}) = -C_\infty + 2\eta\log\sqrt{a^2+b^2}, \tag{11.20}$$

in which C_∞ represents a screening constant. (An overall scale in the argument of the logarithm has again been absorbed into C_∞.) We have chosen this expression as a simple analytic generalization of the Coulomb phase shift for a point charge. This phase shift function is easily shown via the solution to the Abel equation Eq. (2.6a) to correspond to a Coulomb potential

$$V_1(r) = \eta v\left[\frac{1}{a^2+r^2}\right]^{1/2}. \tag{11.21}$$

This potential corresponds to the charge density

$$\rho_1(r) = \frac{3a^2}{4\pi}\left[\frac{1}{a^2 + r^2}\right]^{5/2}, \tag{11.22}$$

which has mean radius

$$\langle r \rangle = 2a. \tag{11.23}$$

The stationary impact parameters associated with the phase shift function Eq. (11.20) are given by the equation

$$X_1'(b_x) = 2\eta \frac{b_x}{a^2 + b_x^2} = q, \tag{11.24}$$

which has only two solutions,

$$b_{jx} = \frac{\eta}{q} \pm \left[\left(\frac{\eta}{q}\right)^2 - a^2\right]^{1/2}. \tag{11.25}$$

These solutions are real inside the Coulomb rainbow point, $q < q_{\mathrm{R}}$, with

$$q_{\mathrm{R}} = \frac{\eta}{a}. \tag{11.26}$$

They are shown in Fig. 11.2 together with the cross section corresponding to $a = 3$ fm, for two different values of η, $\eta = 2$, and $\eta = 10$.

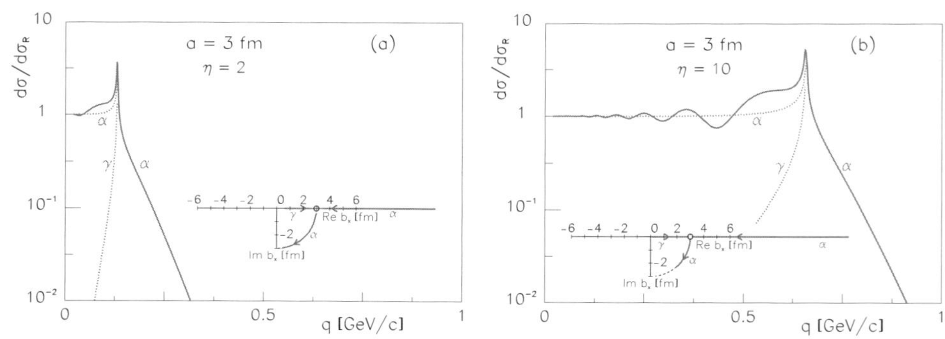

Figure 11.2 Differential cross section ratios $d\sigma/d\sigma_{\mathrm{R}}$ for pure Coulomb scattering from the charge distribution $\rho_1(r) = (3a^2/4\pi)(a^2 + r^2)^{-5/2}$, with $a = 3$ fm. Two values of the Sommerfeld parameter are considered, (a) $\eta = 2$, and (b) $\eta = 10$. The inserts show the trajectories of stationary points for 0.02 GeV/c $\leq q \leq 5.0$ GeV/c (dotted). The parts that correspond to the cross section plots, 0.02 GeV/c $\leq q \leq 1.0$ GeV/c are shown as solid curves. The open circles represent rainbow points. The cross section ratios contributed by the individual stationary points, α and γ, are shown as dotted curves. The solid curves give the resulting interference pattern, with the spikes representing rainbow singularities.

It is instructive to expand Eq. (11.25) inside the Coulomb rainbow point,[1]

$$b_{\alpha x} \simeq \frac{2\eta}{q}, \quad b_{\gamma x} \simeq \frac{\eta q}{2q_R^2}; \quad q \ll q_R. \tag{11.27}$$

The corresponding values of $X_1(b_{jx})$ and $X_1''(b_{jx})$ are approximately

$$X_1(b_{\alpha x}) \simeq 2\eta \log(2\eta/q); \quad X_1(b_{\gamma x}) \simeq 2\eta \log(\eta/q_R); \tag{11.28}$$

$$X_1''(b_{\alpha x}) \simeq -\frac{q^2}{2\eta}; \quad X_1''(b_{\gamma x}) \simeq \frac{2q_R^2}{\eta}. \tag{11.29}$$

The two amplitudes may thus be approximated as

$$F(b_{\alpha x}) = -\frac{2ik\eta}{q^2} \exp\{2i\eta[-1 + \log(2\eta/q)]\}, \tag{11.30a}$$

$$F(b_{\gamma x}) = \frac{k\eta}{2q_R^2} \exp\{2i\eta[-(q/2q_R)^2 + \log(\eta/q_R)]\}. \tag{11.30b}$$

For $q \ll q_R$, $F(b_{\alpha x})$ will dominate $F(b_{\gamma x})$, and the familiar Rutherford cross section is obtained. But, as $q \to q_R$, the interference term becomes increasingly important[2]

$$\begin{aligned} \left(\frac{d\sigma}{dq^2}\right) \Big/ \left(\frac{d\sigma}{dq^2}\right)_{\mathcal{R}} &\simeq 1 + \frac{2\,\mathrm{Re}\,[F(b_{\alpha x})F^*(b_{\gamma x})]}{|F(b_{\alpha x})|^2} \\ &\simeq 1 - \frac{1}{2}\left(\frac{q}{q_R}\right)^2 \sin\left\{2\eta\left[1 - \left(\frac{q}{2q_R}\right)^2 + \log\left(\frac{q}{2q_R}\right)\right]\right\}, \end{aligned} \tag{11.31}$$

valid for $q \ll q_R$.

We rewrite this as

$$1 - \frac{1}{2}\left(\frac{q}{q_R}\right)^2 \sin[2\eta\phi(q)]. \tag{11.32}$$

The period of oscillation will be determined by

$$2\eta[\phi(q_{j+1}) - \phi(q_j)] = \pi, \tag{11.33}$$

or

$$\Delta q = q_{j+1} - q_j \simeq \frac{\pi}{2\eta}\left[\frac{\partial\phi}{\partial q}\right]^{-1}, \tag{11.34}$$

[1] Here, and in the next chapter, we shall denote the dominant stationary point by α, even though it (for Coulomb repulsion) occurs in the fourth quadrant.

[2] We emphasize that "$\mathcal{R}$" stands for "Rutherford," i.e., $(d\sigma/dq^2)_{\mathcal{R}} = (d\sigma/dq^2)_{\text{"Rutherford"}}$, while "R" means "Rainbow," $q_R = q_{\text{"Rainbow"}}$.

where the index j labels the successive minima of the cross section ratio. By using

$$\frac{\partial \phi}{\partial q} = -\frac{q}{2q_R^2} + \frac{1}{q}, \tag{11.35}$$

we obtain

$$\Delta q \simeq \frac{\pi}{2\eta} \frac{q\, q_R^2}{q_R^2 - \frac{1}{2}q^2}. \tag{11.36}$$

Within the present approximation, the period of oscillation thus increases with the momentum transfer, in an almost linear fashion. This behaviour is readily traced back to the phase of $F(b_{\alpha\, x})$, which depends on momentum transfer as $-2\eta\log(q)$. The oscillations are thus due to the rapid variation of the Coulomb phase.

The amplitude of the oscillation increases quadratically with q. This is a reflection of the q^{-2} behaviour of the "outer" Coulomb amplitude, $F(b_{\alpha\, x})$.

The cross sections given in Fig. 11.2 are based on the exact values for the stationary points.[3] The qualitative features of the cross section are seen to be consistent with those derived from the approximation, Eq. (11.27). In particular, the period of oscillation inside the rainbow for $\eta = 10$ changes with q in agreement with Eq. (11.36).

11.4 Gaussian Type Charge Distribution

We next consider a charge distribution which is somewhat less diffuse in nature. A simple such example, which leads to explicit analytic expressions for $\chi(\boldsymbol{b})$, is the generalized Gaussian charge distribution,

$$\rho_2(r) = \frac{1}{\pi^{3/2}\beta^3}\frac{1}{1+\frac{3}{2}\alpha}\left(1+\alpha\frac{r^2}{\beta^2}\right)e^{-r^2/\beta^2}, \tag{11.37}$$

which is often encountered in shell models of the light nuclei. The corresponding Coulomb phase shift function is

$$\chi_2(\boldsymbol{b}) = -C'_\infty + 2\eta\left[\log\sqrt{b^2} + \frac{1}{2}E_1(b^2/\beta^2) + \frac{1}{2}\frac{\alpha}{1+\frac{3}{2}\alpha}e^{-b^2/\beta^2}\right], \tag{11.38}$$

where the exponential integral is defined as [2]

$$E_1(x) \equiv \int_x^\infty \frac{\mathrm{d}z}{z}e^{-z}. \tag{11.39}$$

[3] We do not here compare with the numerical evaluation of the diffraction integral. Because of the long range of the charge distribution, Eq. (11.22), these integrals cannot be evaluated without devising some special technique. A similar problem would of course be encountered if one tried to solve the Schrödinger equation for the potential $V_1(r)$, Eq. (11.21).

For $b_x = 0$, this gives

$$X_2(b_x) = -C'_\infty + 2\eta\left[\log(b_x) + \frac{1}{2}E_1(b_x^2/\beta^2) + \frac{1}{2}\frac{\alpha}{1+\frac{3}{2}\alpha}\,e^{-b_x^2/\beta^2}\right], \tag{11.40}$$

for $\operatorname{Re} b_x > 0$, and

$$X_2(-b_x) = X_2(b_x), \tag{11.41}$$

with

$$X_2'(b_x) = \frac{2\eta}{b_x}\left[1 - \left(1 + \frac{\alpha}{1+\frac{3}{2}\alpha}\frac{b_x^2}{\beta^2}\right)e^{-b_x^2/\beta^2}\right]. \tag{11.42}$$

The stationary point which constitutes a rainbow singularity can be found by solving

$$X_2''(b_x) = -\frac{2\eta}{b_x^2}\left\{1 - \left[1 + \frac{2}{1+\frac{3}{2}\alpha}\left(1 + \alpha + \alpha\frac{b_x^2}{\beta^2}\right)\frac{b_x^2}{\beta^2}\right]e^{-b_x^2/\beta^2}\right\} = 0, \tag{11.43}$$

provided $X_2'(b_x)$ is simultaneously real and positive. For $\alpha = 0$, the solution may be determined as $b_{x,\mathrm{R}} \simeq 1.121\beta$, and thus

$$q_\mathrm{R} = 1.276\frac{\eta}{\beta}; \quad (\alpha = 0). \tag{11.44}$$

The latter relation is qualitatively similar to Eq. (11.26). For a given geometry of the charge distribution, the rainbow value of $b_{x,\mathrm{R}}$ is again proportional to η and inversely proportional to the characteristic radius. As α increases from zero, the rainbow point moves slightly towards smaller momentum transfers.

In Fig. 11.3 we show the ratios of cross sections to the Rutherford cross section and (in the inserted graphs) the trajectories of the stationary points for $\chi_2(\boldsymbol{b})$ corresponding to two values of η ($\eta = 2$, and 10, in Fig. 11.3(a) and 11.3(b),

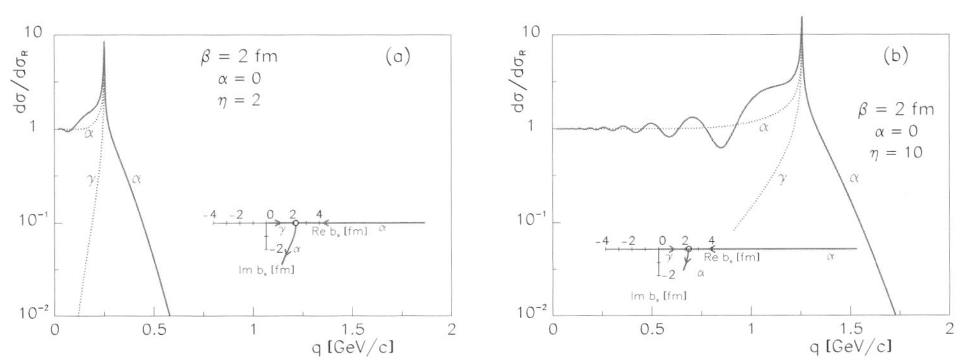

Figure 11.3 As Fig. 11.2, for the generalized Gaussian charge distribution of Eq. (11.37), with $\alpha = 0$ and $\beta = 2$ fm. Trajectories are shown for 0.02 GeV/c $\leq q \leq 2.0$ GeV/c.

respectively), for $\beta = 2$ fm and $\alpha = 0$. The oscillations inside the rainbow point are similar to those found for $\chi_1(\boldsymbol{b})$ in Fig. 11.2. We see in particular, that for this case too, increase of η leads to an oscillating structure and a displacement of the rainbow point to a larger momentum transfer.

An interesting qualitative difference between these two phase shift functions appears outside the rainbow, for $q > q_{\mathrm{R}}$. Whereas the cross section corresponding to $\chi_1(\boldsymbol{b})$ is determined by only one stationary impact parameter, for $\chi_2(\boldsymbol{b})$ it is determined by two stationary impact parameters.

The stationary points $b_{\beta x}$ corresponding to $X_2(b_x)$ for $\alpha = 0$ are far from the real axis and thus not numerically important. However, for non-zero values of α they will move closer to the real axis. In this way, the phase shift function $\chi_2(\boldsymbol{b})$ may lead to cross section ratios that exhibit oscillations outside the rainbow point. This is illustrated in Fig. 11.4 for $\alpha = 2$. (The rainbow point for $\alpha = 2$ is given by $b_{x,\mathrm{R}} = 1.495\beta$ and $q_{\mathrm{R}} = 1.035\eta/\beta$.)

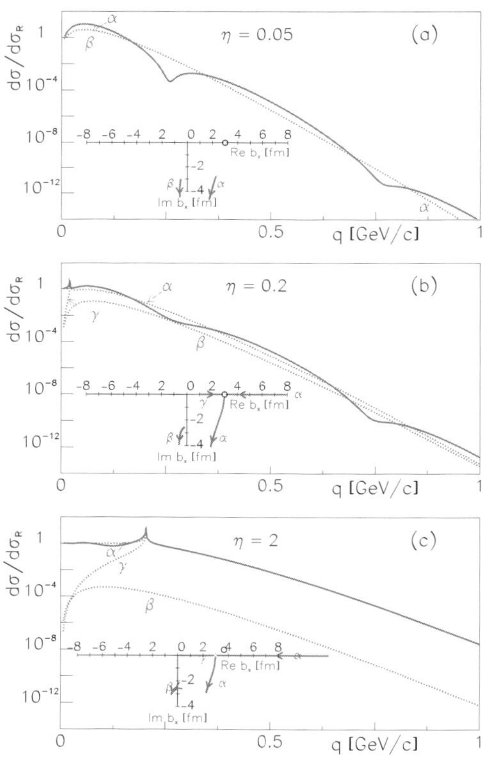

Figure 11.4 Differential cross section ratios $d\sigma/d\sigma_{\mathrm{R}}$ for pure Coulomb scattering from the generalized Gaussian charge distribution of Eq. (11.37), with $\alpha = 2$ and $\beta = 2$ fm. The figure illustrates how for this charge distribution oscillations are present outside the rainbow point for small values of η, but disappear when η becomes large. The trajectories of stationary points shown in the inserts correspond to the range of momentum transfer $0.01\ \mathrm{GeV/c} \leq q \leq 1.0\ \mathrm{GeV/c}$.

Figure 11.4 also illustrates an interesting dependence of the cross section ratio on the value of η. We see that when η is small, oscillations are present, whereas they gradually disappear as η increases. The reason for the disappearance of these oscillations is related to the relative values of the imaginary parts of the phase shifts, $\mathrm{Im}\, X_2(b_{jx})$. It turns out that $\mathrm{Im}\, X_2(b_{\alpha x})$ and $\mathrm{Im}\, X_2(b_{\beta x})$ are quite different in magnitude, and that difference increases rapidly with η. Since $F(b_{jx})$ contains, as a factor, $\exp[-\mathrm{Im}\, X_2(b_{jx})]$, it follows that when η becomes large, one of these amplitudes $(F(b_{\alpha x}))$ will dominate the other one, $(F(b_{\beta x}))$.

11.5 "Fermi" Type Charge Distribution

The third class of charge distributions we consider is the one that corresponds to the thickness function

$$T_C(b) = T_0/\{1 + \exp[(b - c)/\beta]\}^p, \tag{11.45}$$

for $p = 1$ and 2. (The corresponding charge densities [see Eq. (11.7)] can be found by using the appropriate Abel equation.) In this case, the one-dimensional integrals [Eqs. (11.14)–(11.19)] that determine the phase shift function and its derivatives must be evaluated numerically. We denote the phase shift functions corresponding to the above thickness functions as $\chi_3(\boldsymbol{b})$ and $\chi_4(\boldsymbol{b})$, for $p = 1$ and 2, respectively. They are shown in Fig. 11.5 for real values of b_x. At large values, $b_x \gtrsim c$, they are very similar to each other, and to the point-charge phase shift function, $\chi_{\mathrm{pt}}(b_x)$.

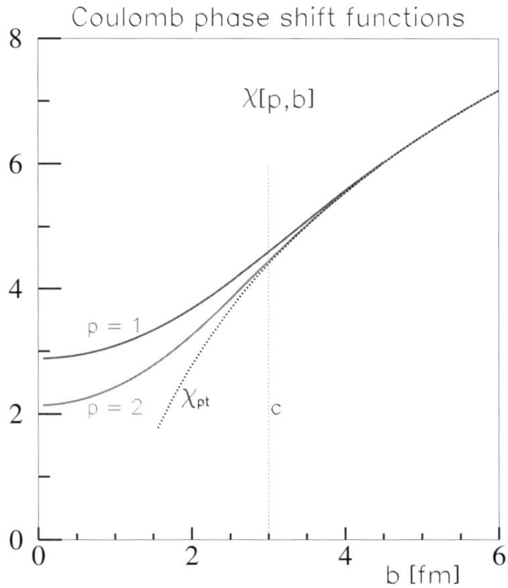

Figure 11.5 Phase shift function $\chi(b)$ corresponding to the thickness function (11.45), for $c = 3$ fm, $\beta = 0.5$ fm, $p = 1$, and $p = 2$. For reference, the phase shift function for a point charge is also shown.

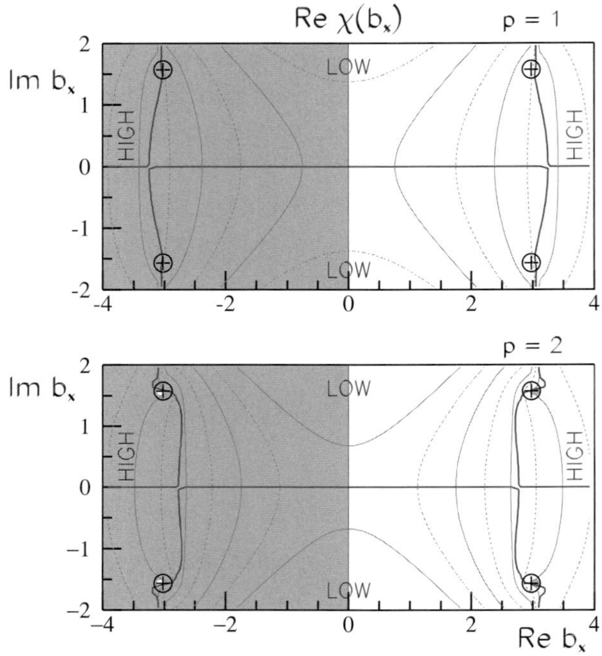

Figure 11.6 Phase shift function $\chi(b)$ corresponding to the thickness function (11.45), for $c = 3$ fm, $\beta = 0.5$ fm, $p = 1$, and $p = 2$, in the complex plane. Singular points at $b_x = \pm(c \pm i\pi\beta)$ are indicated, as well as lines along which $\mathrm{Im}\,\chi(b_x) = 0$ (heavy, dark). Stationary points are possible along these lines, if also $\mathrm{Re}\,\chi(b_x) = 0$ (white regions).

The phase shift functions $\chi_3(b_x)$ and $\chi_4(b_x)$ are explored in more detail in the complex b_x-plane in Fig. 11.6, where contours of constant $\mathrm{Re}\,\chi_3$ and $\mathrm{Re}\,\chi_4$ are shown.

Again, we see that at large values of $|\,\mathrm{Re}\,b_x|$, the two functions are rather similar. Moving away from the real axis, $\mathrm{Re}\,\chi(b_x)$ increases for $|\,\mathrm{Re}\,b_x| > c$, and decreases for $|\,\mathrm{Re}\,b_x| < c$. Actually, there are singularities at $b_x = \pm c + i\pi\beta$ and $b_x = \pm c - i\pi\beta$. Since the phase shift function is defined by an integral along a path in the complex b_x-plane, there will be lines of discontinuity attached to the singular points. These lines of discontinuous $\chi(b_x)$ have been taken parallel to the imaginary axis, away from the real axis.

Also indicated in Fig. 11.6, are regions where $\mathrm{Re}\,\chi'(b_x) > 0$ (white, possible regions of stationary phase points) and $\mathrm{Re}\,\chi'(b_x) < 0$ (dark). An important difference between the two cases is that for $p = 2$ (but not for $p = 1$), there are small regions in the left half-plane where $\mathrm{Re}\,\chi'(b_x) > 0$, indicated in white, close to the singularities $b_x = -c \pm i\pi\beta$. As we will see, the existence of these regions introduces an additional stationary point for $p = 2$.

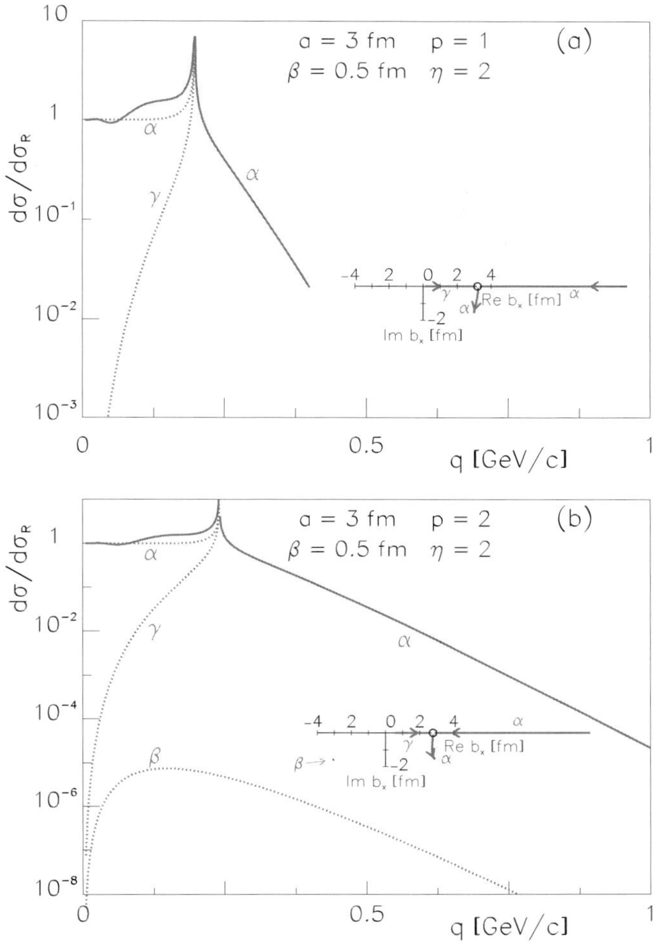

Figure 11.7 Differential cross section ratios $d\sigma/d\sigma_R$ for pure Coulomb scattering from charge distribution corresponding to the thickness function $T_C(b) = T_0/\{1+\exp[(b-c)/\beta]\}^p$, for $c = 3$ fm, $\beta = 2$ fm, and two values of the power p, (a) $p = 1$, and (b) $p = 2$. The strength of the interaction is in both cases the same, $\eta = 2$. As indicated by the inserts, for $p = 1$ there are only two trajectories, α and γ, and hence the differential cross section ratio falls smoothly outside the rainbow point. On the other hand, for $p = 2$, there is also a trajectory of stationary points, β, in the third quadrant, and outside the rainbow point the cross section ratio exhibits oscillations. The trajectories of stationary points shown in the inserts correspond to the range of momentum transfer 0.01 GeV/c $\leq q \leq$ 1.0 GeV/c.

The cross section ratios and the stationary point trajectories corresponding to X_3 and X_4 are shown in Fig. 11.7(a) and 11.7(b), respectively, for $c = 3$ fm, $\beta = 0.5$ fm, and $\eta = 2$. For $p = 1$ (Fig. 11.7(a)), there are two trajectories of stationary impact parameters, denoted $b_{\alpha x}$ and $b_{\gamma x}$, whereas for $p = 2$ (Fig. 11.7(b)), there are three, which we denote $b_{\alpha x}$, $b_{\beta x}$, and $b_{\gamma x}$. As a consequence, the cross

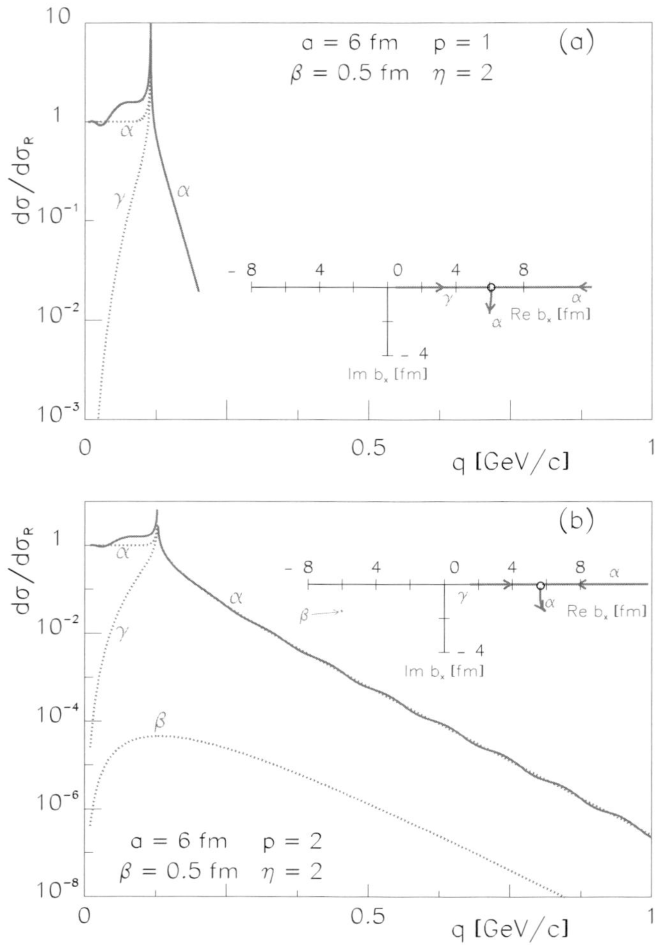

Figure 11.8 As Fig. 11.7, for $c = 6$ fm.

section for $p = 1$ has a smooth behaviour for $q > q_R$, whereas for $p = 2$ interference between $F(b_{\alpha x})$ and $F(b_{\beta x})$ leads to an asymptotically regular oscillatory pattern in the differential cross section, with a period inversely proportional to c. This behaviour is shown in Fig. 11.8, where the corresponding plots have been given for a radius twice as large, $c = 6$ fm.

One interesting property of the phase shift function X_4 ($p = 2$) is that the trajectory of stationary points $b_{\beta x}$ is limited to a very small neighborhood of the singularity, located at $b_x = -c - i\pi\beta$. Also, in contrast to what we found for the Gaussian charge distribution [$\chi_2(\boldsymbol{b})$, with $\alpha \neq 0$], the oscillations of the cross section ratios remain strong, even for large values of η. This behaviour is related to the fact that the stationary points are closer to the real axis, so that $\operatorname{Im} X(b_{\alpha x})$ and $\operatorname{Im} X(b_{\beta x})$ are less different in magnitude.

A general comment can be made here concerning the existence of the trajectory of stationary points, $b_{\beta x}$. Let us suppose the thickness function has a singularity of the type

$$T(b) = T_{\rm R}(b - b')^{-\nu} + \text{L.S.T.}, \tag{11.46}$$

where "L.S.T." is meant to indicate "less singular terms." The stationary points will then be given asymptotically for large q by

$$\epsilon \left[4\pi\eta \frac{T_{\rm R}}{1-\nu}(b_{jx} - b'_x)^{-(\nu-1)} + \text{L.S.T.} \right] = q, \quad \text{for } \nu \neq 1, \tag{11.47}$$

and

$$\epsilon \left[4\pi\eta T_{\rm R} \log(b_{jx} - b'_x) + \text{L.S.T.} \right] = q, \quad \text{for } \nu = 1, \tag{11.48}$$

where

$$\epsilon = \text{Re}\,(b_{jx})/|\,\text{Re}\,(b_{jx})|. \tag{11.49}$$

We can assume without loss of generality that $\eta > 0$. Changing the sign of η simply interchanges the roles of the stationary points α and β.

A simple analysis then shows that Eq. (11.47) will have solutions $b_{\beta x}$ [Re $(b_{\beta x}) < 0$] as well as $b_{\alpha x}$ [Re $(b_{\alpha x}) > 0$], provided $\nu > 3/2$. On the other hand, for $\nu \leq 3/2$, the solution $b_{\beta x}$ does not exist. This is indeed the case for $X_3(b_x)$. It may be noted that the above argument breaks down if the singularity is integrable, i.e., for $\nu < 1$. These results are summarized in Table 11.1.

Our study of Coulomb scattering from extended charge distributions can be summarized as follows: Inside the rainbow point the cross section ratios exhibit oscillations due to the interference of amplitudes contributed by trajectories $b_{\alpha x}$ and $b_{\gamma x}$. The crucial feature leading to these characteristic oscillations is the fact that the secondary amplitude, $F(b_{\gamma x})$, has a phase that varies slowly compared to that of the dominant, "outer" Coulomb amplitude, $F(b_{\alpha x})$. Due to the infinite tail of the Coulomb potential, the rate of change of the phase of $F(b_{\alpha x})$ is, for small q, approximately $2\eta/q$.

Table 11.1 *Existence of $b_{\beta x}$ for $T(b) = T_R(b-b')^{-\nu}$+L.S.T.*

Range in ν	Existence of $b_{\beta x}$
$\nu < 1$	No statement
$1 \leq \nu \leq \frac{3}{2}$	$b_{\beta x}$ does not exist
$\frac{3}{2} < \nu$	$b_{\beta x}$ exists

This may be contrasted with the corresponding rate of change of the amplitudes resulting from the short-range real phase shift functions which were studied in Chapters 7 and 8. For example, for the real Gaussian phase shift function, the rate of change of the phase $-qb_{\alpha x} + X(b_{\alpha x})$ behaves for small q roughly as $\sqrt{\log q}$, as can be seen from Eq. (7.8). In the next chapter we shall see that, as combinations of Coulomb and strong interactions are considered, similar patterns will arise. However, in such cases, the most important among the secondary amplitudes may as well be associated with the nuclear interactions as with the charge distribution (as was the case here).

For $q > q_{\mathrm{R}}$, on the other hand, the cross section may or may not show oscillations, depending on the existence and location of the trajectory $b_{\beta x}$.

Patterns similar to some of those shown in this chapter are observed in electron–nucleus scattering [5], where the asymptotic approach may also be of use. Electron scattering differs from hadron scattering, however, in two important aspects:

- Spin effects are important at large momentum transfers.
- At moderate energies, noneikonal effects, i.e., corrections to elementary diffraction theory are important. These may be accounted for by adding a suitable $1/k$-correction term to the phase shift function.

12

Coulomb Plus Nuclear Interactions

So far, we have discussed phase shift functions which increase or decrease monotonically with the impact parameter b. These are either short-range nuclear interactions whose magnitudes are falling off with b, or long-range Coulomb interactions, slowly increasing with b. In this chapter we will discuss combinations of such interactions. The first of these phase shift functions will be purely real. Again, some qualitatively new features appear.

With combinations of short-range and Coulomb interactions, the overall phase shift function can be written as

$$X(b_x) = X_{\mathrm{N}}(b_x) + X_{\mathrm{C}}(b_x), \tag{12.1}$$

where $X_{\mathrm{N}}(b_x)$ is some short-range nuclear interaction and the Coulomb part has the property that

$$X_{\mathrm{C}}(b_x) \to X_{\mathrm{pt}}(b_x) = -C_\infty + 2\eta \log b_x, \tag{12.2}$$

for b_x much larger than the Coulomb screening radius, and with C_∞ determined by details of the screening. For the Coulomb part, we shall throughout this chapter adopt the phase shift function Eq. (11.20) studied in Section 11.3 with $C_\infty = 0$,

$$X_{\mathrm{C}}(b_x) = 2\eta \log \sqrt{a^2 + b_x^2}. \tag{12.3}$$

Such combinations Eq. (12.1) of phase shift functions can lead to a rich variety of cross section patterns.

Because of the long tail of the Coulomb phase shift function, the exact numerical integral against which we want to compare the asymptotic approximation, needs to be evaluated with some care. This can be achieved by writing

$$\exp[iX(b_x)] = \exp[iX_{\mathrm{C}}(b_x)] + \{\exp[iX(b_x)] - \exp[iX_{\mathrm{C}}(b_x)]\}. \tag{12.4}$$

Then, the diffraction integral can be split as [see Eq. (7.29)]

$$\begin{aligned} F(q) &= -ik \int_0^\infty J_0(qb_x) \exp[iX(b_x)] b_x \, \mathrm{d}b_x \\ &= F_\mathrm{C}(q) - ik \int_0^\infty J_0(qb_x) \{ \exp[iX(b_x)] - \exp[iX_\mathrm{C}(b_x)] \} b_x \, \mathrm{d}b_x \\ &= F_\mathrm{C}(q) + ik \int_0^\infty J_0(qb_x) \exp[iX_\mathrm{C}(b_x)] \{ 1 - \exp[iX_\mathrm{N}(b_x)] \} b_x \, \mathrm{d}b_x . \end{aligned} \tag{12.5}$$

In this last form, it is clear that the unspecified phase arising from screening remains an overall factor, common to both parts. For the Coulomb amplitude $F_\mathrm{C}(q)$ we take the form given in Section 11.3 with $C_\infty = 0$, corresponding to the phase shift function Eq. (12.3). The remaining term, involving the curly bracket in the last line of Eq. (12.5), is a convergent integral.

There is one other technical issue here. With a distributed charge like that represented by the phase shift function Eq. (12.3), the asymptotic approximation to the Coulomb amplitude $F_\mathrm{C}(q)$ that we extract will have a rainbow singularity at some momentum transfer $q_\mathrm{R}^\mathrm{Coul}$. This singularity is an artifact of the asymptotic approximation, there is no analogous singularity contained in the numerical integral against which it could cancel; the overall amplitude has no singularity at that value of the momentum transfer (but possibly at some other value).

12.1 Real Gaussian Plus Coulomb

We consider first the composite phase shift function

$$X(b_x) = A_\mathrm{g} \exp(-b_x^2/\beta^2) + 2\eta \log \sqrt{a^2 + b_x^2} . \tag{12.6}$$

As a reference, we shall keep the Coulomb part fixed, with $\eta > 0$, corresponding to repulsion. For the short-range nuclear part,

$$X_\mathrm{N}(b_x) = A_\mathrm{g} \exp(-b_x^2/\beta^2), \tag{12.7}$$

we shall consider both attraction ($A_\mathrm{g} > 0$) and repulsion ($A_\mathrm{g} < 0$).

For simple phase shift functions, including those that are real and fall off monotonically in magnitude as the impact parameter b increases, there is a simple relationship between the classical concepts of attraction and repulsion and the sign of the phase shift function, as discussed in Chapter 3. We recall that the force is related to the gradient of the potential, whereas the phase shift function is given by a line integral through the potential. For these simple phase shift functions, a

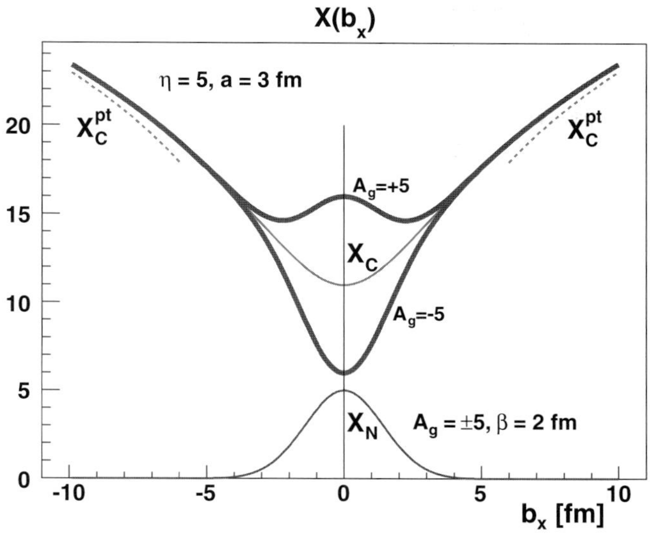

Figure 12.1 Phase shift functions $X_C(b_x)$ and $X_N(b_x)$, with X_N given as the Gaussian of Eq. (12.7), as well as their sum (for both signs of A_g). For comparison, also the point-Coulomb phase shift function is indicated at large $|b_x|$ as X_C^{pt}.

repulsive potential ($V(r) > 0$) leads to a negative phase shift function ($\chi(b) < 0$) whose derivative is positive ($X'(b_x) > 0$ for b_x real and positive). The Coulomb phase shift function is superficially in conflict with this correspondence, since we have subtracted the infinite, negative screening constant, C_∞.

For a representative set of parameters, we show the two parts of this phase shift function, as well as the sum (for both signs of A_g) in Fig. 12.1. The corresponding derivatives, which for a given momentum transfer q determine the points of stationary phase, are given in Fig. 12.2. Again, we display both the individual contributions and their sum (for both signs of A_g). Clearly, at small momentum transfers, the Coulomb part will dominate, with the main contribution coming from the far-out Coulomb stationary point, and the scattering cross section will be similar to those described in Section 11.3.

However, there will be additional contributions from one or more stationary points at shorter distances. We see from Fig. 12.2 that if the short-range interaction is repulsive, with $A_g < 0$, there will be one additional stationary point, as was already the case for the distributed-charge phase shift function $\chi_1(b)$ studied in Section 11.3. But now the additional stationary point is occurring at a shorter distance than was the case for the pure Coulomb scattering, and the corresponding rainbow point will occur at a higher value of momentum transfer. For an attractive short-range interaction, however, there can at small momentum transfers be a total

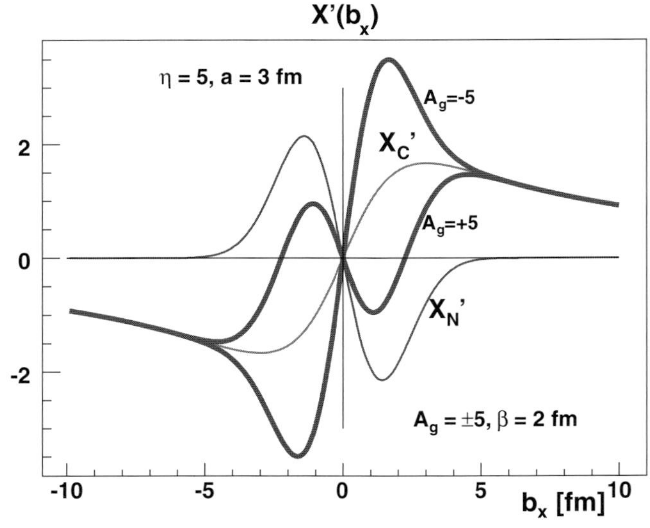

Figure 12.2 Derivatives of phase shift functions: $X'_{\mathrm{C}}(b_x)$ and $X'_{\mathrm{N}}(b_x)$, with X_{N} given as the Gaussian of Eq. (12.7), as well as their sum (for both signs of A_{g}).

of four points of stationary phase, as indeed is the case for the parameters adopted for Figs. 12.1 and 12.2. This will lead to *two* rainbow points! In contrast to the secondary rainbow observed in the scattering of light from water drops, here the two rainbows are quite independent of each other, occurring at rather different scattering angles. In fact, one rainbow is due to repulsion (positive $\mathrm{Re}\, b_x$), the other is due to attraction ($\mathrm{Re}\, b_x < 0$).

On the other hand, at large momentum transfers the short-range interaction will dominate, the points of stationary phase will occur off the real axis, giving a differential cross section reminiscent of those described for the real Gaussian phase shift function in Sections 7.1 and 7.3.

In analogy with Fig. 7.2, we show in Figs. 12.3 and 12.4 for the two cases $A_{\mathrm{g}} < 0$ and $A_{\mathrm{g}} > 0$, respectively, regions in the complex b_x-plane where $\mathrm{Re}\, X'(b_x)$ is positive (white) and where it is negative (shaded). Stationary points will be found where $\mathrm{Re}\, X'(b_x)$ is positive, and where simultaneously $\mathrm{Im}\, X'(b_x)$ vanishes (dashed lines superimposed on the figure, including in these cases the real axis). Also, where two such lines of vanishing $\mathrm{Im}\, X'(b_x)$ cross at 90° (in a white region), there will be a rainbow point.

When dealing with more complicated phase shift functions, such "maps" are very useful for easily identifying where possible stationary points can be found (along dashed lines in the white regions). They also reveal where rainbow points are to be found (where two dashed curves cross at an angle of 90°).

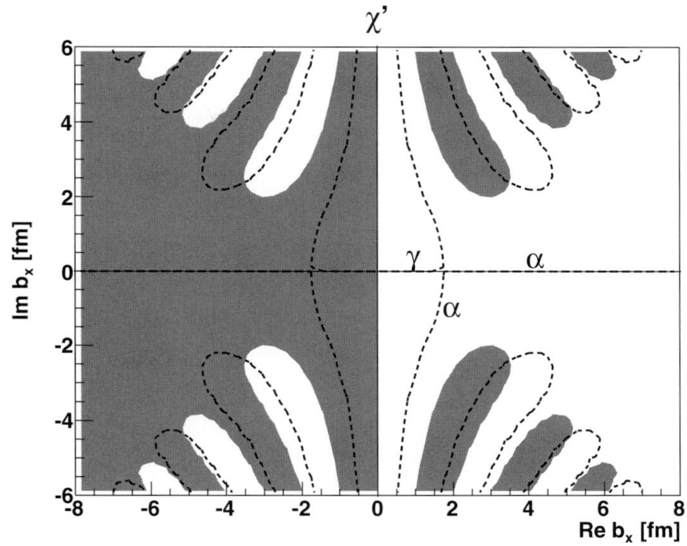

Figure 12.3 Regions of positive (white) and negative (shaded) values of $\mathrm{Re}\, X'(b_x)$, shown in the complex b_x-plane for $A_g = -5$. The figure is analogous to Fig. 7.2, except that here $\beta = 2$ fm and that the phase-shift function now also has a Coulomb part. Also indicated as dashed lines, are the locations where $\mathrm{Im}\, X'(b_x) = 0$ (in this case, including the real axis). A rainbow point occurs at $b_x = 1.65$ fm.

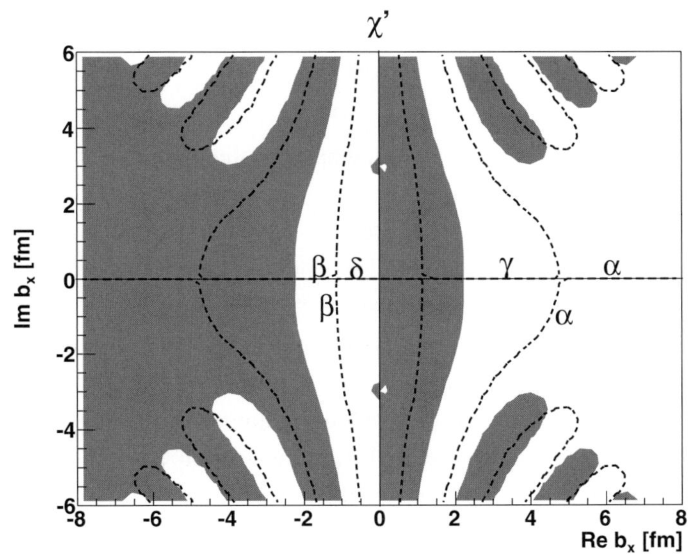

Figure 12.4 Analogous to Fig. 12.3 except that the short-range interaction is also attractive, $A_g = +5$. Rainbow points occur for $b_{R1} = -1.09$ fm and $b_{R2} = 4.55$ fm.

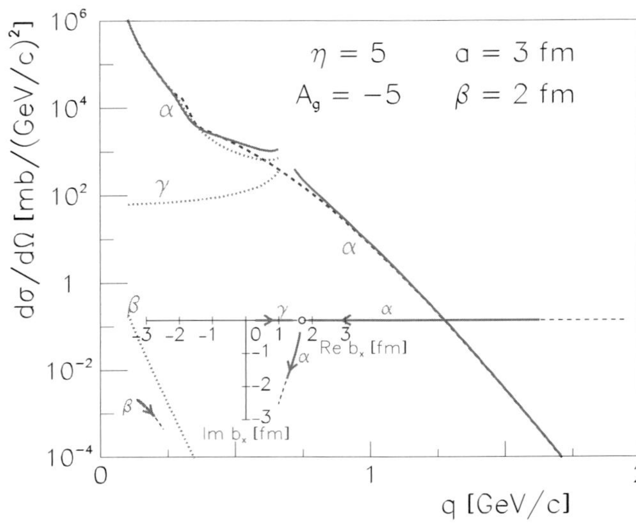

Figure 12.5 Differential cross section for the phase shift function Eq. (12.6). The inserts show the trajectories of stationary points for $0.16\ \text{GeV/c} \leq q \leq 6.0\ \text{GeV/c}$ (dotted). The parts that correspond to the cross section plots, $0.2\ \text{GeV/c} \leq q \leq 2.0\ \text{GeV/c}$ are shown as solid curves. The open circle represents the rainbow point. The cross sections contributed by the individual stationary points, α, β, and γ, are shown as dotted curves. The solid curves give the resulting interference pattern, with the gap around $q_R = 0.69$ GeV/c representing a rainbow singularity.

12.1.1 Repulsive Gaussian Plus Coulomb

We start by discussing the case of a repulsive short-range interaction. As is seen from Figs. 12.2 and 12.3, this is a somewhat simpler case than the mixed case, in the sense that there is only one rainbow.

In Fig. 12.5 we display the differential cross section corresponding to the combined phase shift function Eq. (12.6) for this case of short-range repulsion, $A_g = -5$. A forward Coulomb cross section (completely dominated by the contribution of the Coulomb stationary-phase point α) is followed by a transition region with a rainbow at somewhat larger momentum transfer than in the pure Coulomb case of Section 11.3, determined by the contributions of the stationary points α and γ. Then, at larger values of q, there is a smooth fall-off, similar to that seen in Fig. 7.3 for the real Gaussian phase-shift function. The contribution from the stationary point β is everywhere subdominant.

The agreement between the asymptotic approach and the numerical result (shown by the dashed curve) is again quite good, except for the region of the rainbow point ($q_R = 0.69$ GeV/c), as well as the region of the Coulomb rainbow point ($q_R^{\text{Coul}} = 0.33$ GeV/c), as discussed above. Immediately beyond the rainbow point, the cross section is seen to fall more rapidly. This can be understood in terms of the stationary point α moving away from the real axis.

12.1.2 Attractive Gaussian Plus Coulomb

As was already evident from Fig. 12.2, when the short-range interaction is attractive ($A_{\rm g} > 0$), the situation is more rich in structure, with the possibility of having two rainbow points. For the parameters chosen for that figure, there are indeed two rainbow points. Whether this phase shift function leads to one or two rainbows, is determined by whether or not $X'(b_x = 0)$ has a positive or negative slope at the origin. In fact, for the phase shift function Eq. (12.6), we have

$$X'(b_x) \to -2\frac{b_x}{\beta^2}A_{\rm g} + 2\eta\frac{b_x}{a^2}, \quad \text{as } b_x \to 0. \tag{12.8}$$

With $\eta > 0$, we see that this slope becomes negative (see Fig. 12.2) for $A_{\rm g} > 0$ (attraction) *and* provided

$$\eta < \left(\frac{a}{\beta}\right)^2 A_{\rm g}, \tag{12.9}$$

which then, within this model, is the condition for having two rainbows. For the parameters adopted in Figs. 12.1, 12.2, and 12.4 these two rainbow points are located at $q_{\rm R1} = 0.19$ GeV/c and $q_{\rm R2} = 0.29$ GeV/c.

The resulting differential cross section is shown in Fig. 12.6. At low momentum transfers, the asymptotic approach has contributions from four points of stationary phase (see also Fig. 12.2). Two of these are at positive values of $\mathrm{Re}\, b_x$ (the far-out one is associated with the Coulomb repulsion), and two are at negative values. The latter are associated with the short-range attraction.

The two rainbow points, which are rather close to each other, make it natural to discuss separately three regions of momentum transfer. At low momentum transfers, inside the first rainbow (at $q_{\rm R1} = 0.19$ GeV/c, corresponding to the nuclear attraction), the cross section is dominated by the outer Coulomb trajectory, denoted α. In fact, the other contributions, including those corresponding to the first rainbow, β and δ, are subdominant compared to the α-contribution. As a result, this first rainbow singularity is practically invisible.

Between the two rainbows, there are three contributions, from the stationary points denoted α, β, and γ. At the second rainbow ($q_{\rm R2} = 0.29$ GeV/c), the points of stationary phase α and γ run together, and consequently the γ contribution will here be comparable to that coming from the trajectory denoted α.

Immediately beyond the second rainbow, at $q_{\rm R}^{\rm Coul} = 0.33$ GeV/c, there is the (unphysical) Coulomb rainbow associated with the subtracted amplitude $F_{\rm C}(q)$.

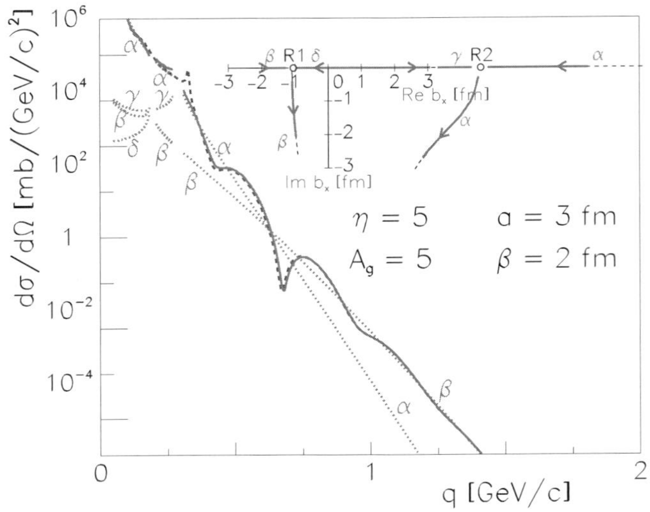

Figure 12.6 Differential cross section for the phase shift function Eq. (12.6). The inserts show the trajectories of stationary points for $0.1\ \text{GeV/c} \leq q \leq 5.0\ \text{GeV/c}$ (dotted). The parts that correspond to $0.1\ \text{GeV/c} \leq q \leq 2.0\ \text{GeV/c}$ are shown as solid curves ($0.22\ \text{GeV/c} \leq q$ for α). The two open circles labeled "R1" and "R2" represent rainbow points. The cross sections contributed by the individual stationary points, α, β, γ, and δ, are shown as dotted curves. The solid curves give the resulting interference pattern, with gaps around $q_{R1} = 0.19$ GeV/c and $q_{R2} = 0.29$ GeV/c representing rainbow singularities.

Further out in momentum transfer, the contributions of α and β cross over in magnitude, leading to oscillations in the range of momentum transfer where they are comparable. Here, the stationary points α and β are both off the real axis. In contrast to the previous case of a repulsive nuclear interaction, in this case it is the β trajectory that dominates at large momentum transfers, not the "Coulomb" trajectory α.

12.2 Absorptive Gaussian Plus Coulomb

As an example of nuclear absorption plus Coulomb attraction, we now turn to the phase shift function Eq. (12.6) with $A_g = 5i$. The possible locations of stationary points are again easily determined from the map of real and imaginary parts of $X'(b_x)$ shown in Fig. 12.7, similar to those for the real cases, Figs. 12.3 and 12.4.

Again, there is a "Coulomb" trajectory (denoted α), which at small momentum transfers comes arbitrarily close to the real axis, but which for finite momentum transfers is always below the real axis. For the case of Coulomb repulsion considered here, it is located in the fourth quadrant of the complex b_x-plane. The cases of Coulomb repulsion and attraction are related by the symmetry Eq. (2.10).

The corresponding differential cross section is shown in Fig. 12.8. The result of a direct numerical integration is shown as a dashed curve, and is seen to yield a poor

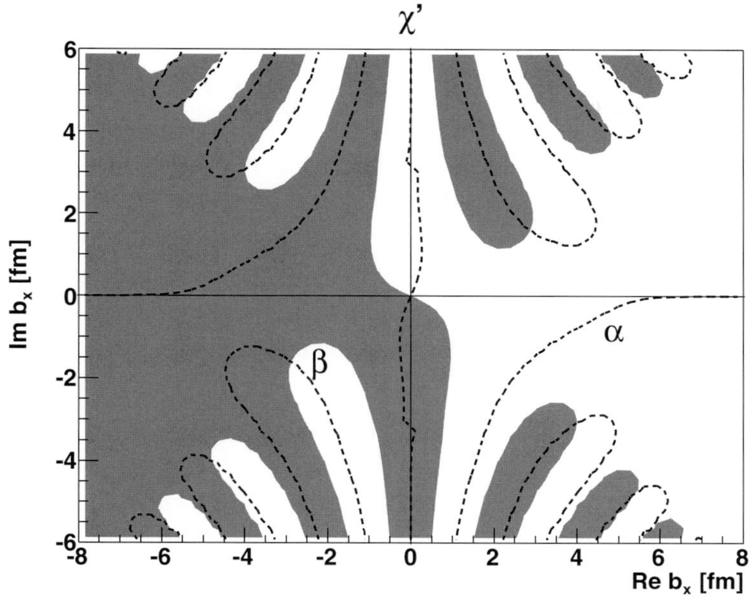

Figure 12.7 Regions of positive and negative values of Re $X'(b_x)$, shown in the complex b_x-plane for $A_g = 5i$. The figure is analogous to Figs. 12.3 and 12.4, except that here the short-range nuclear interaction is purely absorptive.

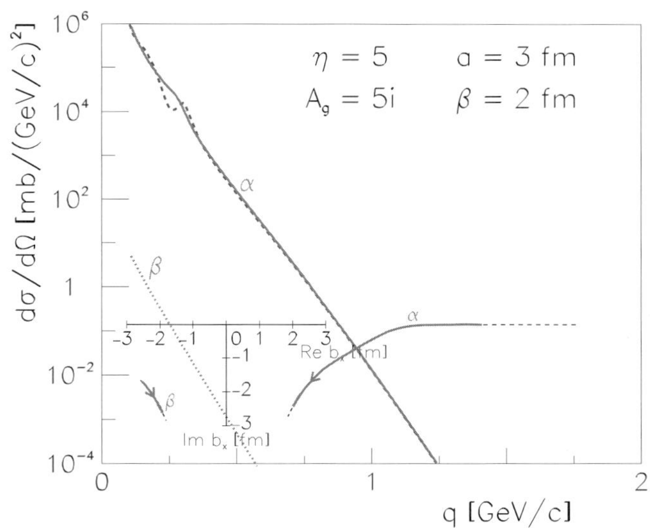

Figure 12.8 Differential cross section for the phase shift function Eq. (12.6). The inserts show the trajectories of stationary points for 0.15 GeV/c $\leq q \leq$ 5.0 GeV/c (dotted). The parts that correspond to the cross section plots, 0.2 GeV/c $\leq q \leq$ 3.0 GeV/c are shown as solid curves. The cross sections contributed by the individual stationary points, α and β are shown as dotted curves. The solid curve gives the resulting interference pattern.

result in the region of the Coulomb rainbow $q_R \simeq 0.33$ GeV/c, as discussed in the context of Eq. (12.5). As compared with the case of a purely absorptive Gaussian phase shift function, but no additional Coulomb interaction, studied in Section 7.3 (see, in particular, Fig. 7.6), the present phase shift function yields a very smooth cross section, with no oscillations. In terms of the trajectories of stationary points, this is caused by the α-trajectory being quite a bit closer to the real axis in the complex b_x-plane, due to the considerable strength of the Coulomb field. In fact, were the strength of the Coulomb field to be reduced, oscillations would gradually appear. At the same time, the envelope would become more steep. The case of a weaker Coulomb field will be illustrated in the next section.

12.3 "Fermi" Plus Coulomb

Consider next the phase shift function

$$X(b_x) = \frac{A_F}{1+\exp[(b_x - c)/\beta]} + 2\eta \log\sqrt{a^2+b_x^2}, \qquad \mathrm{Re}\, b_x > 0,$$
$$X(-b_x) = X(b_x). \tag{12.10}$$

The short-range part of this has been studied in Section 8.4, under the title "Fermi" phase shift function. It has two shape parameters, c and β, for the radius and diffuseness of a "Fermi" distribution. The Coulomb part is the same as that considered previously in this chapter.

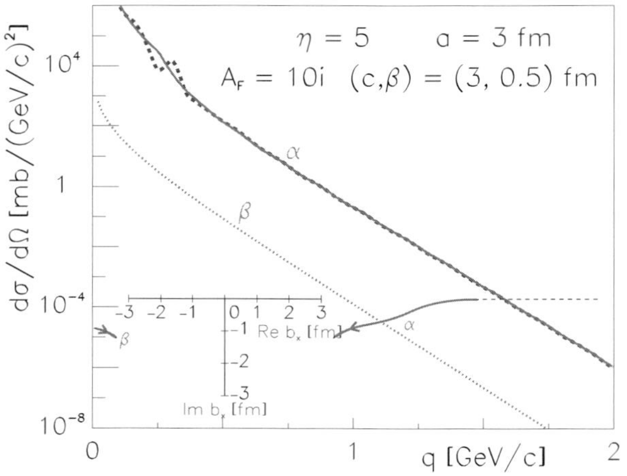

Figure 12.9 Differential cross section for the phase shift function Eq. (12.10). The inserts show the trajectories of stationary points for 0.15 GeV/c $\leq q \leq 4.0$ GeV/c (dotted). The parts that correspond to 0.2 GeV/c $\leq q \leq 3.0$ GeV/c are shown as solid curves. The cross sections contributed by the individual stationary points, α and β are shown as dotted curves. The solid curve gives the resulting interference pattern. The heavy dashed curve gives the result of a numerical integration.

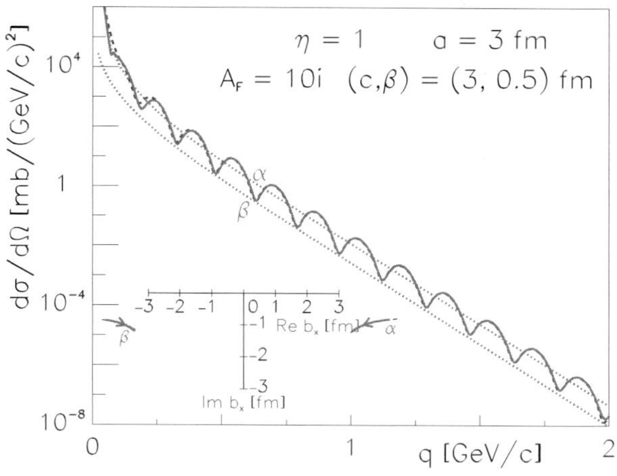

Figure 12.10 Differential cross section for the phase shift function Eq. (12.10). The inserts show the trajectories of stationary points for 0.15 GeV/c $\leq q \leq$ 4.0 GeV/c (dotted). The parts that correspond to 0.2 GeV/c $\leq q \leq$ 3.0 GeV/c are shown as solid curves. The cross sections contributed by the individual stationary points, α and β are shown as dotted curves. The solid curve gives the resulting interference pattern. The heavy dashed curve gives the result of a numerical integration.

We here focus on the case of an absorptive short-range interaction, with A_F pure imaginary. The shape parameters c and β are taken as 3 fm and 0.5 fm, respectively, values also studied in Section 8.4. For the strength of the Coulomb interaction, we consider two values, first a rather strong field, $\eta = 5$, in Fig. 12.9, and then a more moderate value, $\eta = 1$, in Fig. 12.10.

For the case of a stronger Coulomb field, the differential cross section is quite smooth. After a forward peak dominated by the Coulomb interaction, it has a smooth falloff, similar to that seen in Fig. 12.8 for the same strength of the Coulomb interaction. It is at all values of momentum transfer dominated by the contribution from the trajectory denoted α, which is the continuation of the Coulomb trajectory "down" into the absorptive regime, i.e., away from the real axis.

In Fig. 12.10 we study the case of a weaker Coulomb field, $\eta = 1$. This is roughly the magnitude relevant to proton scattering from lead. Compared with the case of a strong Coulomb field, the relative contributions from trajectories α and β are here more comparable, and, as a result, the differential cross section exhibits pronounced oscillations, similar to those seen in Section 8.4. The minima are however not zeros, since the amplitudes contributed by the two stationary points are somewhat different in magnitude, and therefore cannot completely cancel.

13

Two-Scale Phase Shift Functions

Phase shift functions can of course also have a rich structure without the inclusion of a long-range Coulomb interaction. The present chapter presents two such examples. The first example is simply a superposition of two Gaussians, whereas the second example is defined in terms of the derivative $X'(b_x)$ rather than the phase shift function $X(b_x)$ itself.

13.1 Two Gaussians

We shall here consider the superposition of two concentric Gaussians having however different ranges, $\beta_1 \neq \beta_2$:

$$X(b_x) = A_1 \exp(-b_x^2/\beta_1^2) + A_2 \exp(-b_x^2/\beta_2^2). \tag{13.1}$$

As two explicit examples, we consider

$$X_1(b_x) = 5 \exp[-(b_x/\beta_1)^2] + 3 \exp[-(b_x/\beta_2)^2], \tag{13.2a}$$

$$X_2(b_x) = 5 \exp[-(b_x/\beta_1)^2] - 3 \exp[-(b_x/\beta_2)^2], \tag{13.2b}$$

with $\beta_1 = 1$ fm and $\beta_2 = 2$ fm. These phase shift functions, and their first derivatives, are shown in Figs. 13.1 and 13.2, respectively. The first form, Eq. (13.2a), is attractive throughout, whereas the second form, Eq. (13.2b) has a long-range part corresponding to repulsion.

The corresponding maps of regions in the complex b_x-plane, where $\operatorname{Re} X'(b_x)$ is respectively positive and negative, are shown in Figs. 13.3 and 13.4 for the two phase shift functions. Together with the contours where $\operatorname{Im} X'(b_x) = 0$, these reveal where points of stationary phase may occur. For the present phase shift functions, these maps have a very rich and complicated pattern in the region close to the imaginary axis. This is due to the interplay of the two Gaussians, each of which becomes singular for $b_x \to \pm i \cdot \infty$.

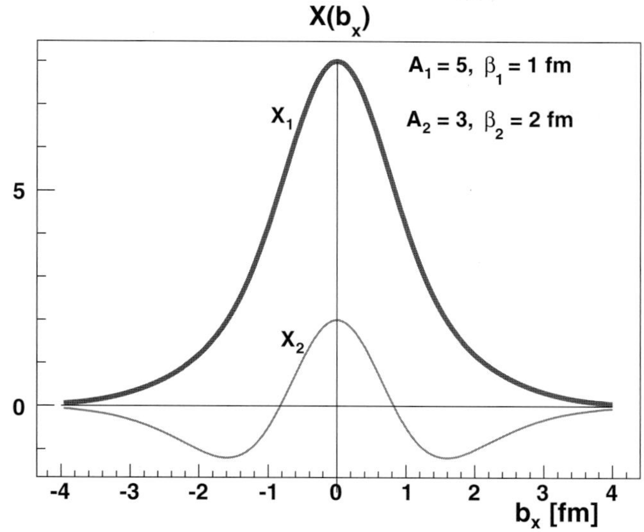

Figure 13.1 The double-Gaussian phase shift functions Eq. (13.2).

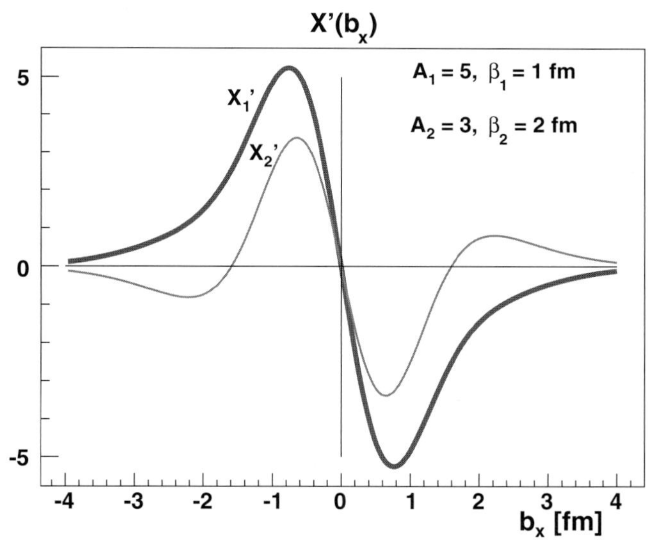

Figure 13.2 Derivatives of the double-Gaussian phase shift functions Eq. (13.2).

We shall discuss in some detail the second of these phase shift functions, Eq. (13.2b), which leads to two rainbow points and in that sense has a richer structure. For the parameters adopted in Figs. 13.1, 13.2, and 13.4, the rainbow points occur at ($b_{\mathrm{R1}} = 2.22$ fm, $q_{\mathrm{R1}} = 0.16$ GeV/c) and ($b_{\mathrm{R2}} = -0.65$ fm, $q_{\mathrm{R2}} = 0.67$ GeV/c). The lower rainbow point, at positive b_x, can be associated with the

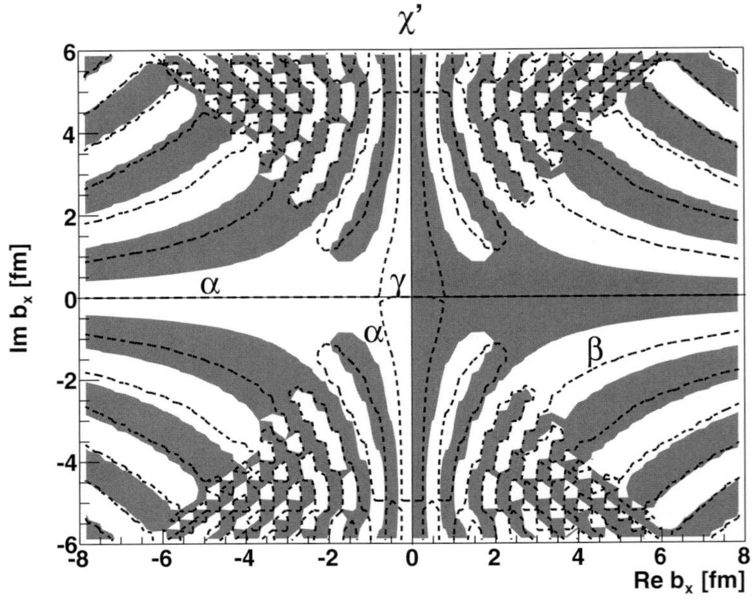

Figure 13.3 Regions of positive (white) and negative (shaded) values of Re $X'(b_x)$, shown in the complex b_x-plane for the double-Gaussian phase shift function Eq. (13.2a).

long-range repulsion, whereas the higher one, at negative b_x, is to a large extent determined by the short-range attractive part of the interaction. It has an analogue at a similar value of b_R for the all-attractive phase shift function Eq. (13.2a).

The following question arises: For this case of a rainbow singularity at $b_x = b_{R2} < 0$, how does the path of integration enter from $\text{Re}\, b_x = -\infty$? In particular, is there any additional point of stationary phase further out, as appears quite possible from the map shown in Fig. 13.4? In other words, does the valley coming in from $\text{Re}\, b_x = -\infty$ pass through the shaded region (in which case there is no further point of stationary phase) or not? This question can easily be answered by studying $|\exp[-iqb_x + iX(b_x)]|$ in the complex b_x plane, a further point of stationary phase is encountered. The corresponding point of stationary phase will be referred to as ϵ, but its contribution to the scattering amplitude is not quantitatively significant.

The differential cross section resulting from this phase shift function Eq. (13.2b) has in the region of the rainbow singularities a rich structure, as illustrated in Fig. 13.5. Inside the first rainbow point, $q < q_{R1}$, there are contributions from four points of stationary phase on the real axis, denoted as α, β, γ, and δ, plus one below the real axis, in the third quadrant, denoted ϵ. The contribution from β (at large positive b_x, corresponding to the repulsive interaction) is the dominant one. At the rainbow point, it diverges, as does the contribution from δ.

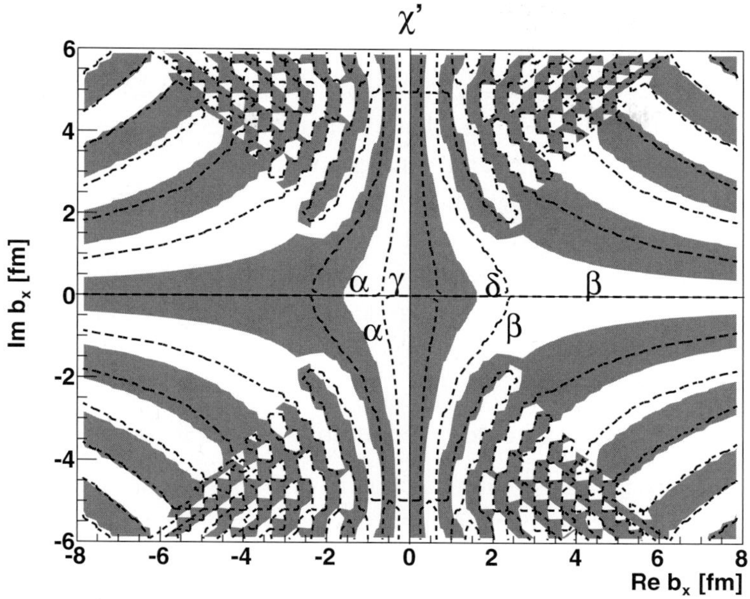

Figure 13.4 Regions of positive (white) and negative (shaded) values of Re X' (b_x), shown in the complex b_x-plane for the double-Gaussian phase shift function Eq. (13.2b).

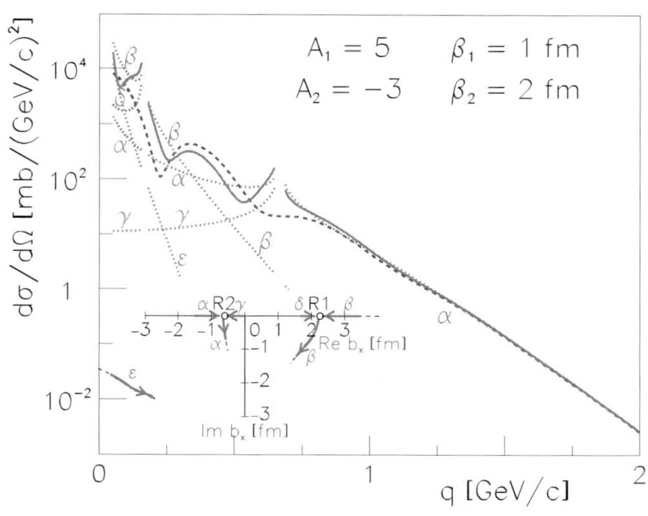

Figure 13.5 Differential cross section for the phase shift function Eq. (13.2b). The inserts show the trajectories of stationary points for $0.02\ \mathrm{GeV/c} \leq q \leq 5.0\ \mathrm{GeV/c}$ (dotted). The parts that correspond to the cross section plots, $0.05\ \mathrm{GeV/c} \leq q \leq 2.0\ \mathrm{GeV/c}$ are shown as solid curves. The cross sections contributed by the individual stationary points, are shown as dotted curves. The solid curves give the resulting interference pattern. The heavy dashed curve represents the result of direct numerical integration.

Then, between the two rainbows, there are three contributions. In the first part of this interval, it is the β-contribution which continues to dominate, while in the outer part of the interval, as $q \to q_{R2}$, the contributions from α and γ dominate and eventually become singular. An oscillatory pattern develops, due to these different contributions having different dependencies on q.

Outside the second rainbow, the differential cross section has a smooth falloff, with contributions from β and α, the latter being dominant. Again we see that in the region of the rainbows, the agreement is not very good. As discussed above, both the asymptotic approach and the numerical one have difficulties here, for different reasons.

13.2 The "Shoulder Function"—Three Rainbows

If the derivative of the phase shift function has two positive maxima, $X'(b_x) > 0$, there will be two rainbow points, as we have seen illustrated in various cases. It is interesting to study what happens when also the minimum between these two maxima is *positive*. It is easy to find a phase shift function with this property, if we simply start with $X'(b_x)$, rather than $X(b_x)$. Thus, in analogy with the case studied in Section 10.2.1, we let

$$f(b_x) \equiv -(b_x/\beta_1^2)A_1 \exp(-b_x^2/\beta_1^2) - (b_x^3/\beta_2^4)A_2 \exp(-b_x^2/\beta_2^2), \tag{13.3}$$

and take

$$X'(b_x) = f(b_x). \tag{13.4}$$

For β_1 somewhat smaller than β_2, this function $f(b_x)$ has one maximum controlled mostly by A_1 and the first Gaussian, and a second one further out, in part because of β_2 being larger than β_1, and in part because of the higher power of b_x in front. An even higher (but odd) power could also be considered.

An example is shown in Fig. 13.6, for $A_1 = A_2 = 5$, $\beta_1 = 1$ fm and $\beta_2 = 3$ fm. It is clear that at low momentum transfers there will be two points of the stationary phase on the real axis. As the momentum transfer is increased, at some point q_{R1} there will be a first rainbow. But this rainbow differs from those we have encountered so far in having a *larger* number of stationary points as the momentum transfer is further increased. Immediately beyond this value of momentum transfer, there are four points of stationary phase. Thus, it is clear that for $q < q_{R1}$, there must be three points of stationary phase, one of them off the real axis, in the lower half-plane. With the parameters chosen for Fig. 13.6, the rainbow points occur at $q_{R1} = 0.32$ GeV/c, $q_{R2} = 0.54$ GeV/c and $q_{R3} = 0.86$ GeV/c,

As the momentum transfer is further increased, the remaining two rainbows are encountered, with pairs of stationary points running together at q_{R2} and q_{R3}. This will be illustrated shortly.

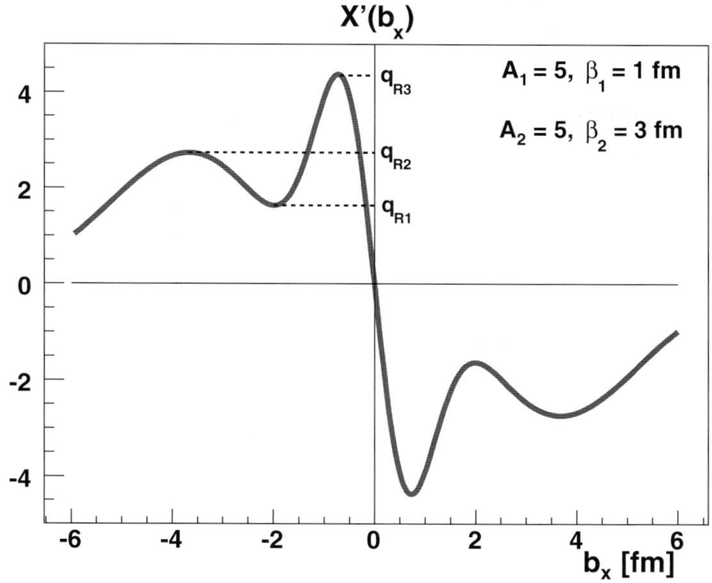

Figure 13.6 The derivative, Eq. (13.4), of a phase shift function yielding three rainbow points, denoted $q_{\rm R1}$, $q_{\rm R2}$, and $q_{\rm R3}$. As the momentum transfer is increased from zero, the number of stationary points along the real axis is seen to go from two to four to two to none. There will be additional points of stationary phase in the complex plane: first one, then two.

In order to calculate the cross section, we need also the phase shift function itself, obtained by integration:

$$\begin{aligned} X(b_x) &= -\int_{b_x}^{\infty} \mathrm{d}t\, X'(t) = -\int_{b_x}^{\infty} \mathrm{d}t\, f(t), \qquad \mathrm{Re}\, b_x > 0, \\ X(-b_x) &= X(b_x). \end{aligned} \tag{13.5}$$

This procedure is completely analogous to the case studied in Section 10.2.1. The result is shown in Fig. 13.7 for real values of b_x. We shall refer to this function as a "shoulder function."

The differential cross section, shown in Fig. 13.8, is in this case not so well reproduced by the asymptotic approximation. This is actually to be expected, due to the fact that the rainbow singularities are relatively near each other. Approximating the contribution of each point of stationary phase by a Gaussian integral from minus infinity to plus infinity then becomes inaccurate. However, there is an interesting new feature. The lowest-lying rainbow point, at $(b_{\rm R1}, q_{\rm R1})$, and labeled "R1" in the figure, has the property that for $q < q_{\rm R1}$, there is only one contribution associated with it, from the point of stationary phase γ, located off the real axis, in the lower half-plane. Beyond it, for $q > q_{\rm R1}$, there are two contributions, from the points of stationary phase γ and δ, both on the real axis.

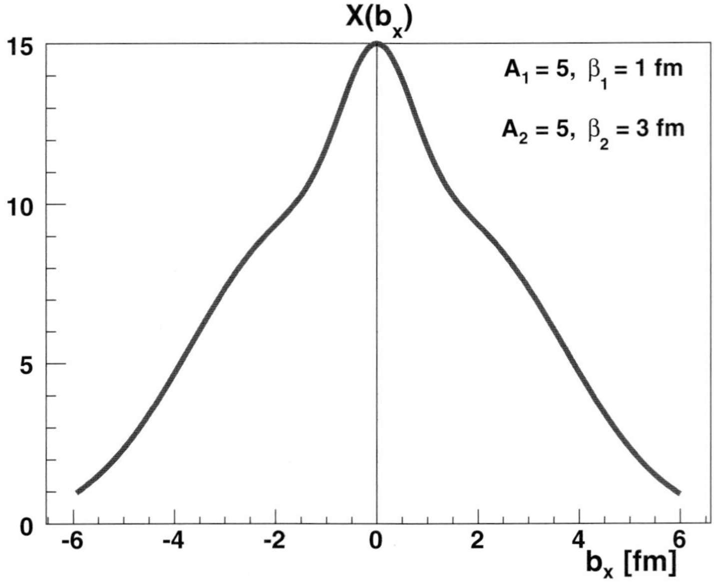

Figure 13.7 The phase shift function corresponding to the derivative given by Eq. (13.4).

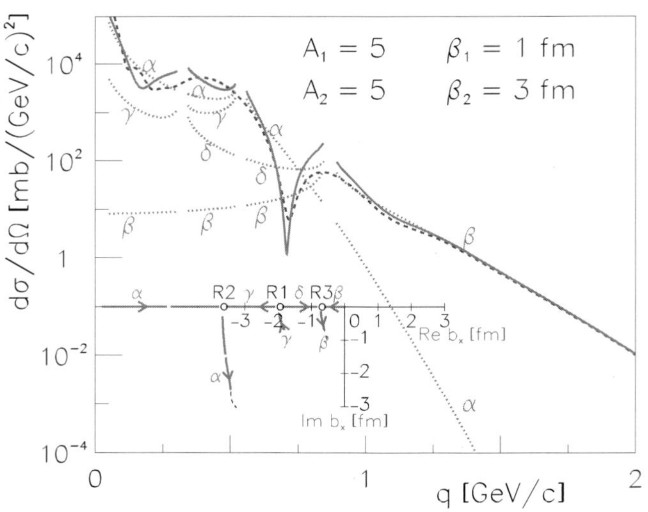

Figure 13.8 Differential cross sections corresponding to the real phase shift function whose derivative is given by Eq. (13.4).

For increasing values of the momentum transfer, but in the region $q_{R1} < q < q_{R2}$, these two points of stationary phase, γ and δ, move away from the rainbow point b_{R1}. The next rainbow point occurs at (b_{R2}, q_{R2}), where γ meets α and there ceases to exist, while α then moves into the complex plane. Likewise, δ meets β at

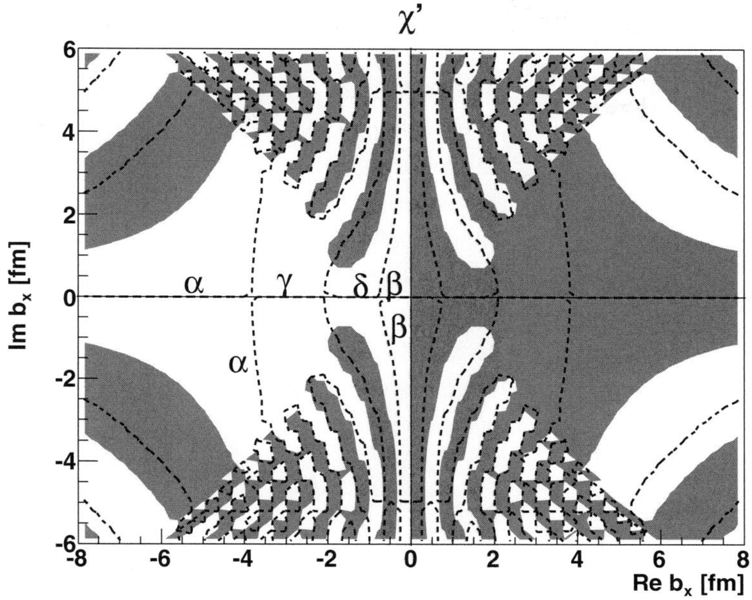

Figure 13.9 Regions of positive and negative values of $\mathrm{Re}\, X'(b_x)$, shown in the complex b_x-plane for the phase shift function whose derivative is given by Eq. (13.4).

the rainbow point $(b_{\mathrm{R3}}, q_{\mathrm{R3}})$ and there ceases to exist. Thus, the point of stationary phase δ only exists for a finite range of momentum transfers.

These possible locations of the stationary points can also be seen from the map in Fig. 13.9. This map has a lot of structure around the imaginary axis, this is due to the Gaussians becoming singular as $\mathrm{Im}\, b_x \to \pm\infty$, and the modulation with a "high" power of b_x.

Another unfamiliar property of the trajectory of stationary points denoted γ becomes clear from an inspection of Fig. 13.9: Whereas trajectories for forward scattering typically start at $\pm\infty$ or 0, the trajectory γ starts at a finite point in the complex b_x-plane, namely where the contour indicating $\mathrm{Im}\, X'(b_x) = 0$ intersects the borderline between the allowed white region (where $\mathrm{Re}\, X'(b_x) > 0$) and the forbidden shaded region (where $\mathrm{Re}\, X'(b_x) < 0$). This is thus a point where both real and imaginary parts of $X'(b_x)$ vanish. It is however not a rainbow point since there are no crossing lines corresponding to $\mathrm{Im}\, X'(b_x) = 0$.

13.3 The Shoulder Function for Absorption

We shall here discuss the absorptive version of the phase shift function defined by its derivative, Eq. (13.4). As we saw for the real case, in Section 13.2, this

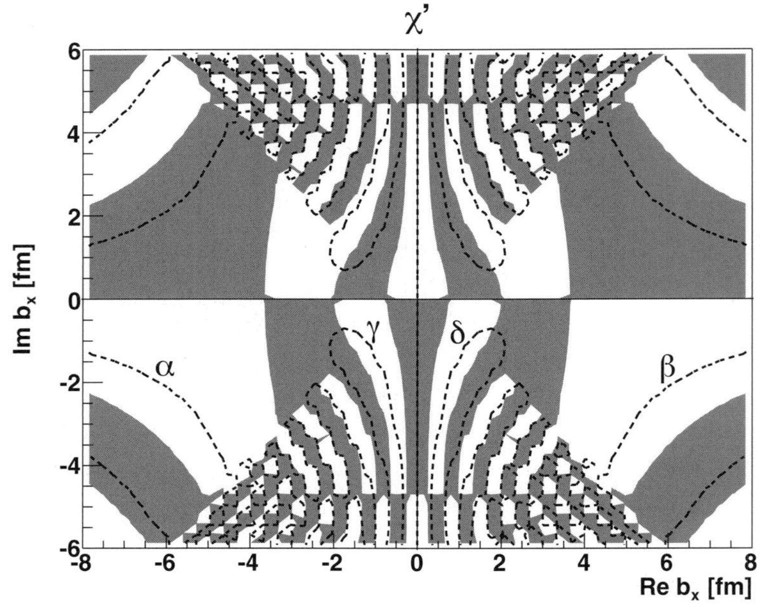

Figure 13.10 Regions of positive and negative values of Re $X'(b_x)$, shown in the complex b_x-plane for the purely imaginary phase shift function whose derivative is given by Eq. (13.4).

functional form can yield a differential cross section with a very rich structure. In the purely imaginary case, as shown in the "map" of Fig. 13.10, there are two pairs of trajectories, one pair associated with each of the two length scales present in Eq. (13.4). As is the case for all purely imaginary phase shift functions, the two points of stationary phase are in each pair related by reflection about the imaginary b_x-axis. In isolation each such pair would thus lead to an oscillatory differential cross section, with an exponentially falling envelope, where the period and slope of the envelope would be determined by the real and imaginary parts of b_x.

Adopting the geometry parameters (β_1 and β_2) of Section 13.2, these two pairs would separately lead to oscillations of rather different periods. This is indeed what happens, as is seen in Fig. 13.11, where we display the cross section out to somewhat larger momentum transfer than was done for the real case.

At moderate momentum transfers, there is a regular pattern resulting from the "outer" points of stationary phase, α and β, mostly determined by the Gaussian with the larger value β_2. The contributions from the stationary points γ and δ are subdominant, since those points are initially further away from the real axis. Eventually, however, these two points of stationary phase, α and β, move further away from the real axis, the contributions from γ and δ become dominant, and the period of oscillation changes dramatically, as does the slope of the envelope.

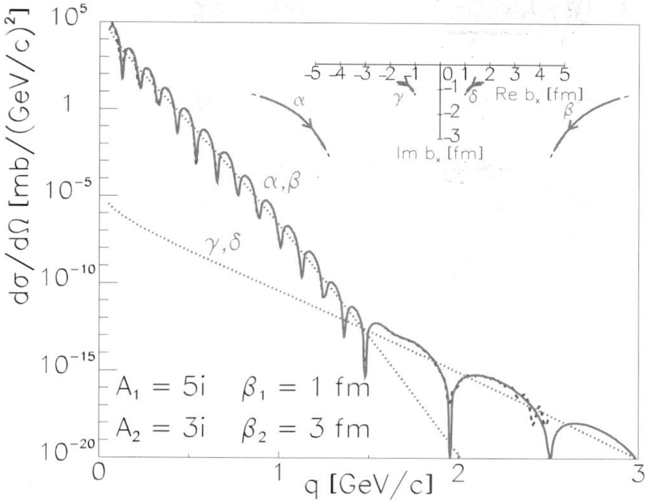

Figure 13.11 Differential cross sections corresponding to the phase shift function whose derivative is given by Eq. (13.4). The inserts show the trajectories of stationary points for 0.02 GeV/c $\leq q \leq$ 5.0 GeV/c (dotted). The parts that correspond to the cross section plots, 0.05 GeV/c $\leq q \leq$ 3.0 GeV/c are shown as solid curves. The cross sections contributed by the individual stationary points, are shown as dotted curves. The solid curve gives the resulting interference pattern. The heavy dashed curve represents the result of the numerical diffraction integral.

In this case, the differential cross section falls by 25 orders of magnitude, and the numerical integration (shown as a heavy dashed curve, practically covered by the solid curve) eventually becomes inaccurate. It is only shown out to $q = 2.5$ GeV/c.

Calculations of differential cross sections such as the one we have just presented raise a difficult problem for experimental measurements. We have noted already in Fig. 1.1 that the cross section for proton collisions with ^{208}Pb shows a consistently periodic variation while decreasing over a range of some seven orders of magnitude. It might be natural to assume that extending the measurements to still smaller scattering intensities would reveal nothing new. The cross section in Fig. 13.11, on the other hand, drops by some 17 orders of magnitude before it begins to behave in a radically different way. The drop need not be that dramatic. Altered behavior may intervene at any level.

14

Double "Fermi" Phase Shift Functions

We shall here consider the superposition of two "Fermi" functions having different ranges, $c \neq c'$, and different surface thicknesses, $\beta \neq \beta'$. Thus, the two phase shift functions that we combine have the same functional form, and their derivatives have singularities of the same degree. Asymptotically, they will thus "attract" the points of stationary phase with equal strength, but the numerical values of the various parameters will determine which singularities are more relevant.

If we denote the basic "Fermi" distribution

$$X_{\mathrm{F}}(c, \beta, b_x) = \frac{A_{\mathrm{F}}}{1 + \exp[(b_x - c)/\beta]}, \tag{14.1}$$

then one could imagine combining such functions either in a multiplicative way

$$[X_{\mathrm{F}}(c, \beta, b_x) + 1]\, X_{\mathrm{F}}(c', \beta', b_x), \tag{14.2}$$

or, in an additive way, as

$$X_{\mathrm{F}}(c, \beta, b_x) + X_{\mathrm{F}}(c', \beta', b_x). \tag{14.3}$$

We shall in this chapter explore this second option:

$$X(b_x) = \frac{A_{\mathrm{F}}}{1 + \exp[(b_x - c)/\beta]} + \frac{A'_{\mathrm{F}}}{1 + \exp[(b_x - c')/\beta']}, \qquad \mathrm{Re}\, b_x > 0, \tag{14.4}$$

with A_{F} and A'_{F} complex and $X(-b_x) = X(b_x)$. In the context of absorption, various cases come to mind: the second function can be exploited to provide more (or less) absorption in the inner region,

$$|A_{\mathrm{F}} + A'_{\mathrm{F}}| > |A_{\mathrm{F}}| \quad \text{or} \quad |A_{\mathrm{F}} + A'_{\mathrm{F}}| < |A_{\mathrm{F}}|, \quad c' < c. \tag{14.5}$$

Another possibility would be to let the real and absorptive parts have different ranges.

Even within this family of functions Eq. (14.4), there are many ways of choosing A'_{F}, c', and β' relative to A_{F}, c, and β. We first explore the case $\beta = \beta'$, and in a second section let β be different from β'.

14.1 Two "Fermi" Phase Shift Functions with the Same Surface Thickness

14.1.1 Additional Internal Structure

We start out by studying the two cases

$$X_p(b_x) = \frac{A_{\mathrm{F}}}{1+\exp[(b_x-c)/\beta]} + \frac{A'_{\mathrm{F}}}{1+\exp[(b_x-c')/\beta]}, \qquad \operatorname{Re} b_x > 0, \quad (14.6a)$$

$$X_m(b_x) = \frac{A_{\mathrm{F}}}{1+\exp[(b_x-c)/\beta]} - \frac{A'_{\mathrm{F}}}{1+\exp[(b_x-c')/\beta]}, \qquad \operatorname{Re} b_x > 0, \quad (14.6b)$$

for purely imaginary phase shift functions with parameters

$$A_{\mathrm{F}} = i, \quad c = 3 \text{ fm}, \quad A'_{\mathrm{F}} = 0.2i, \quad c' = 2 \text{ fm}, \quad \beta = 0.5 \text{ fm}. \qquad (14.7)$$

These phase shift functions are presented in Fig. 14.1 for $A'_{\mathrm{F}} = 0.2i$ (dashed) with $A'_{\mathrm{F}} = 0$ (solid) as a reference.

In order to get an overview over possible locations of stationary points, we start out by showing in Figs. 14.2 and 14.3 the corresponding "maps" of possible stationary points. These are along lines where $X'(b_x)$ is real ($\operatorname{Im} X'(b_x) = 0$, dashed), and where $\operatorname{Re} X'(b_x) > 0$ (white regions). As for the "Fermi" phase shift function discussed in Chapter 8, there are trajectories of stationary points in the regions

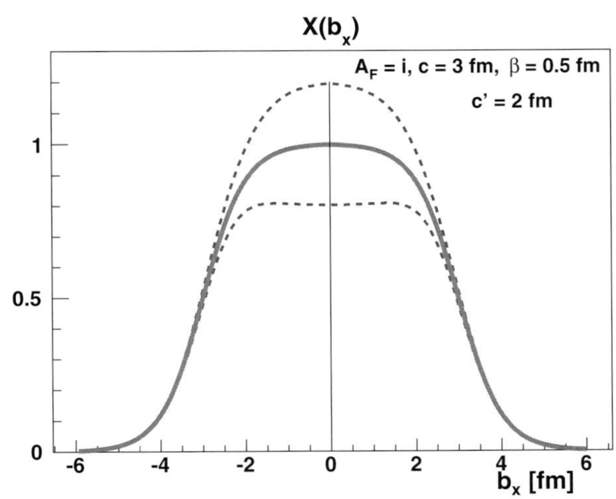

Figure 14.1 Dashed: sum and difference of two "Fermi" phase shift functions having different ranges, but the same surface thickness. Solid: Single "Fermi" phase shift function (reference).

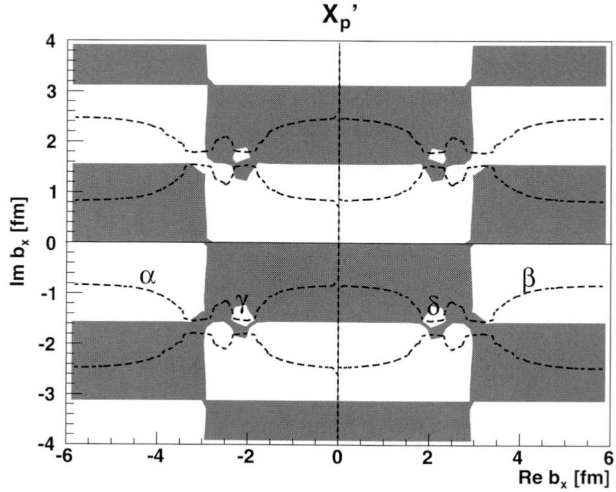

Figure 14.2 Regions of positive and negative values of $\mathrm{Re}\,X'(b_x)$, shown in the complex b_x-plane for the sum of two "Fermi" phase shift functions Eq. (14.6a), with the parameters of Eq. (14.7).

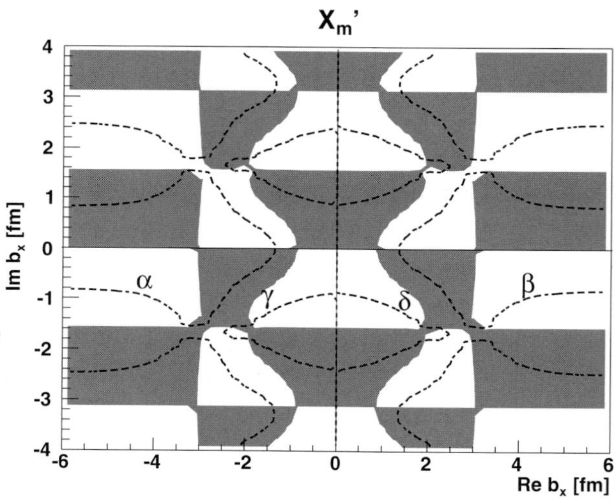

Figure 14.3 Regions of positive and negative values of $\mathrm{Re}\,X'(b_x)$, shown in the complex b_x-plane for the difference of two "Fermi" phase shift functions Eq. (14.6b), with the parameters of Eq. (14.7).

(white rectangles) $|\,\mathrm{Re}\,b_x| > c$, $-\pi\beta/2 < \mathrm{Im}\,b_x < 0$. These provide the dominant contributions to the cross sections, and are labeled α and β. Asymptotically, as $q \to \infty$, they approach the pole singularities, $\mp c - i\pi\beta$. In addition, due to the structure provided by the second "Fermi" distribution, there are stationary points γ and δ near $\mp c' - i\pi\beta$.

We recall that we seek a deformation of the contour of integration, from $b_x = -\infty$ to $b_x = +\infty$, down into the complex plane ($\operatorname{Im} b_x < 0$) such that the integral can be associated with the contributions from a few points of stationary phase, or saddle points. For large values of the momentum transfer q, these will be in the neighborhoods of the singularities of $X'(b_x)$.

Let us consider the situation in the fourth quadrant, where the composite phase shift function $X(b_x)$ has simple poles at $b_x = c - i\pi\beta$ and $b_x = c' - i\pi\beta$. The derivative, $X'(b_x)$ will have double poles at these locations. If we expand b_x around one such point as $b_x - b_x^{\text{pole}} = \rho^{i\theta}$, the derivative will for A_{F} imaginary behave as $X' \sim (i/\rho^2)e^{-2i\theta}$. In order to match the (real) momentum transfer, this quantity has to be real at the stationary point,

$$\operatorname{Im}\left(ie^{-2i\theta}\right) = 0, \quad \text{or} \quad \theta = \frac{\pi}{4} + \frac{n\pi}{2}. \tag{14.8}$$

Among these four directions, indicated by dashed curves in Figs. 14.2 and 14.3, X' is positive at every other, namely at $\theta = \pi/4$ and $\theta = 5\pi/4$. But, as for the single "Fermi" function studied in Section 8.4, the path of integration passes through only one of them (here denoted β). In the language of saddle points, the other is behind the "ridge" and will not be encountered along the path of integration. Likewise, γ and δ are on the side of the corresponding poles that is closer to the real axis.

For the case Eq. (14.6b), there will be a different approach to the singularities at $\mp c' - i\pi\beta$ due to the overall minus sign associated with this "secondary" "Fermi" function. In the language of Eq. (14.8), in the fourth quadrant the relevant X' is positive at $\theta = -\pi/4$ and $\theta = 3\pi/4$. Again, the points γ and δ are on the side of the corresponding poles that is closer to the real axis.

Fig. 14.4 is devoted to the differential cross sections produced by the phase shift function X_p of Eq. (14.6). It is shown together with the corresponding one for only one "Fermi" phase shift function, $A'_{\text{F}} = 0$. There is an almost complete overlap. Actually, the X_m case is also similar. In this particular case of X_p, we have checked that a numerical evaluation of the diffraction integral agrees well with the asymptotic result for the differential cross section from the maximum around 10^4 mb/(GeV/c) at small momentum transfers down to a magnitude level of around 10^{-17} mb/(GeV/c).

The agreement between the cross sections with $A'_{\text{F}} = \pm 0.2i$ and $A'_{\text{F}} = 0$ is partly due to the smallness of the contributions from the additional points of stationary phase, γ and δ, and perhaps even more to the very slow movement of the stationary points along the dominant trajectories α and β. The trajectories are shown in the inset in Fig. 14.4, and in Fig. 14.5 we show β and δ (whose real parts are positive) in more detail. The dominant trajectory (in the fourth quadrant) is practically the same for these three cases, both for the real part and for the imaginary part. For the δ

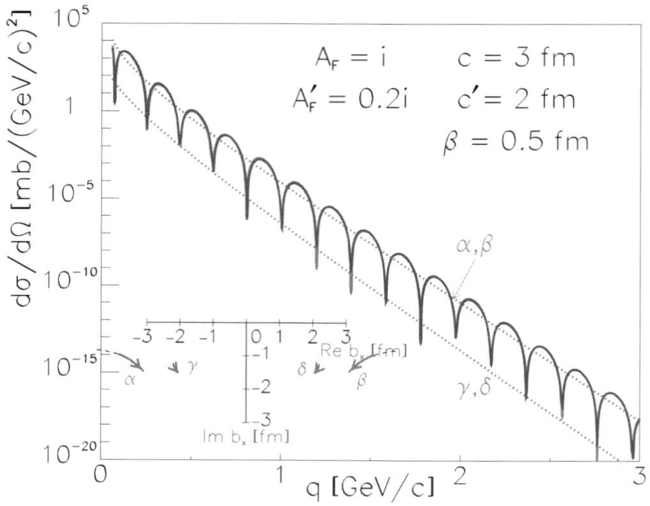

Figure 14.4 Differential cross sections corresponding to the phase shift function Eq. (14.6). The cases of $A_{\mathrm{F}} = 0.2i$ and $A_{\mathrm{F}} = 0$ are completely overlapping. The inserts show the trajectories of stationary points for 0.02 GeV/c $\leq q \leq 5.0$ GeV/c (dotted). The cross sections contributed by the individual stationary points are shown as dotted curves, labeled α, β and γ, δ.

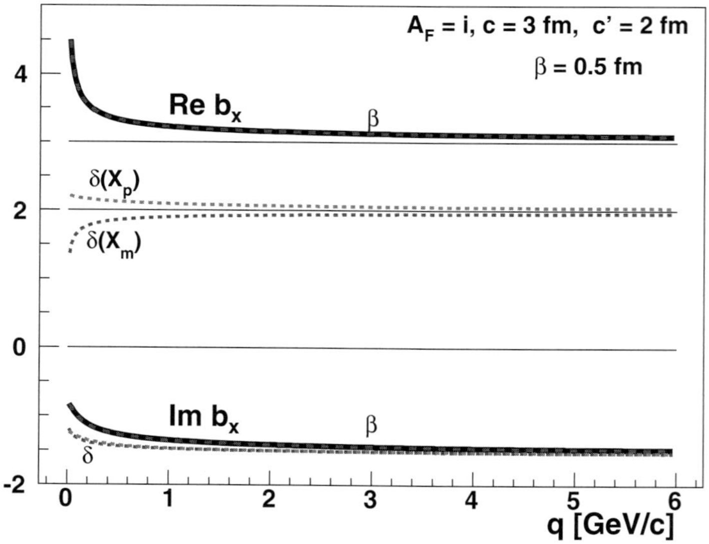

Figure 14.5 Real and imaginary parts of b_x at stationary points β and δ, for a range of momentum transfer q.

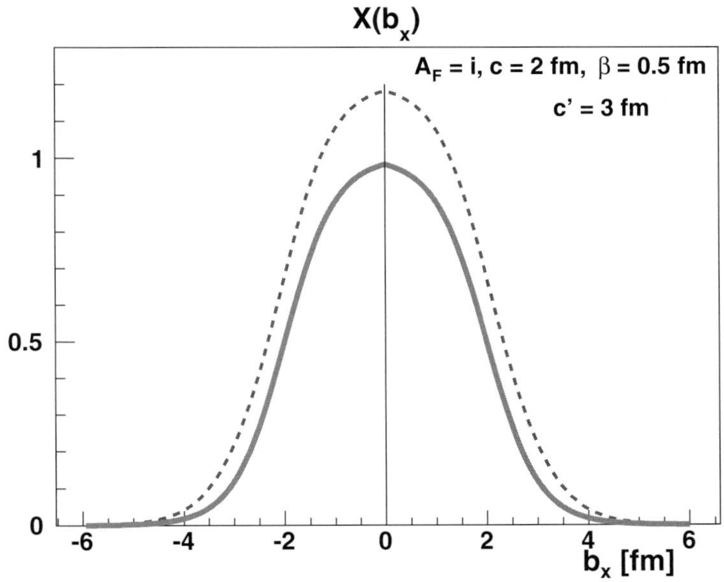

Figure 14.6 "Fermi" phase shift functions. Solid: Single reference function ($A'_\mathrm{F} = 0$). Dashed: sum of two "Fermi" phase shift functions having different ranges, but the same surface thickness, with parameters given by Eq. (14.9).

trajectory, there is in the real part, at small momentum transfers, a small difference between the two cases of X_p and X_m, whereas the imaginary parts are essentially the same.

14.1.2 Additional External Structure

What happens if the outer distribution is subdominant, in the sense that $|A'_\mathrm{F}| < |A_\mathrm{F}|$ with $c' > c$? We display in Fig. 14.6 the phase shift function for such a case, with

$$A_\mathrm{F} = i, \quad c = 2 \text{ fm}, \quad A'_\mathrm{F} = 0.2i, \quad c' = 3 \text{ fm}, \quad \beta = 0.5 \text{ fm}, \tag{14.9}$$

compared with the case of $A'_\mathrm{F} = 0$. The two phase shift functions look qualitatively rather similar, but the double one has an extra pair of stationary points, which for this case are most important.

In Fig. 14.7 we show the "map" of possible stationary points, for the composite phase shift function defined by the parameters of Eq. (14.9). The outer ones, asymptotically approaching the singularities at $b_x = \mp c' - i\pi\beta$ are labeled α and β, whereas the inner ones near $b_x = \mp c - i\pi\beta$ are labeled γ and δ. The corresponding single phase shift function, for $A'_\mathrm{F} = 0$, will have the trajectories α and β connected to the regions of γ and δ.

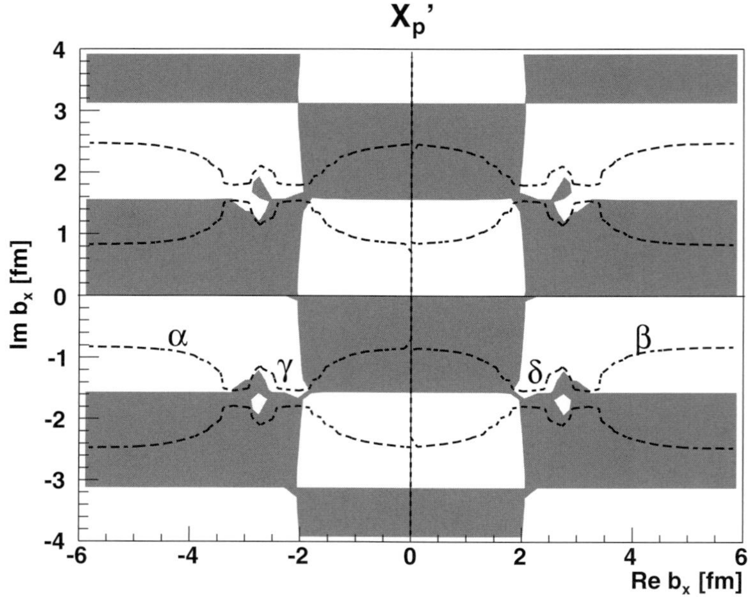

Figure 14.7 Regions of positive and negative values of Re $X'(b_x)$, shown in the complex b_x-plane for the sum of two "Fermi" phase shift functions Eq. (14.6a), with the parameters of Eq. (14.9).

At small momentum transfers, the outer stationary points (α and β) will provide the dominant contributions, since they are closer to the real axis. However, as the momentum transfer increases, these will gradually be forced further down into the complex plane, "attracted" by the external singularities at $\mp c - i\pi\beta$. Then, in some range of momentum transfer, where the two pairs of stationary points (α, β) and (γ, δ) are at similar distances from the real axis, they will provide comparable contributions to the overall scattering amplitude. However, the pair (γ, δ) is closer to each other than the pair (α, β), and therefore leads to oscillations of a longer period.

In Fig. 14.8 we show the resulting differential cross section (solid), and compare with the case of $A'_{\rm F} = 0$ (dashed). While the single "Fermi" phase shift function provides a rather periodic diffraction pattern, the composite one yields at intermediate momentum transfers a more "irregular" pattern. At small momentum transfers it provides a somewhat shorter period of oscillations, corresponding to the contributions of the outer stationary points (α, β) which there (at small momentum transfers) are important. Eventually, as the momentum transfer increases, these contributions are no longer dominant, and those from the inner points of stationary phase, (γ, δ) take over, yielding a longer period of oscillation.

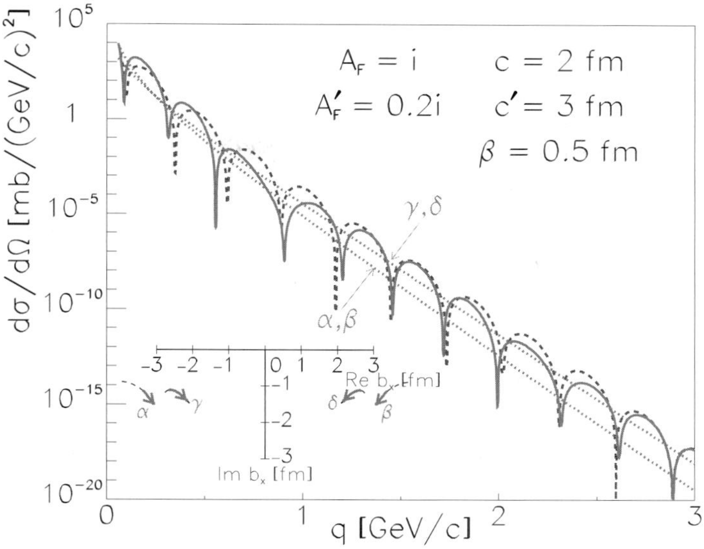

Figure 14.8 Differential cross sections corresponding to the phase shift function Eq. (14.6). The cases of $A'_{\mathrm{F}} = 0.2i$ (solid) and $A'_{\mathrm{F}} = 0$ (dashed) are compared. The inserts show the trajectories of stationary points for $0.02\ \mathrm{GeV/c} \leq q \leq 5.0\ \mathrm{GeV/c}$ (dotted). The cross sections contributed by the individual stationary points are shown as dotted curves, labeled α, β and γ, δ.

Thus, while the differential cross section shown in Fig. 14.4 displays 15 maxima over the range of momentum transfers out to $q = 3$ GeV/c, Fig. 14.8 displays only 10. It is of course no accident that this ratio is that of c'/c.

14.2 Different Surface Thicknesses

The picture can change qualitatively if the two phase shift functions have rather different surface thicknesses. We shall discuss in turn the two cases of an additional external sharper structure, and an internal one, as compared to the "main" phase shift function, with $|A_{\mathrm{F}}| = 1$.

14.2.1 Sharper External Structure

As a first case, let us consider an additional external structure ($c' > c$) which is sharper ($\beta' < \beta$) than the dominant one ($|A_{\mathrm{F}}| = 5|A'_{\mathrm{F}}|$):

$$\begin{aligned} A_{\mathrm{F}} &= i \qquad & c &= 3\ \mathrm{fm} \qquad & \beta &= 0.50\ \mathrm{fm} \\ A'_{\mathrm{F}} &= 0.2i \qquad & c' &= 3.1\ \mathrm{fm} \qquad & \beta' &= 0.25\ \mathrm{fm} \end{aligned} \tag{14.10}$$

The corresponding "map" of possible stationary points is shown in Fig. 14.9. The main part of the phase shift function provides the trajectories of stationary

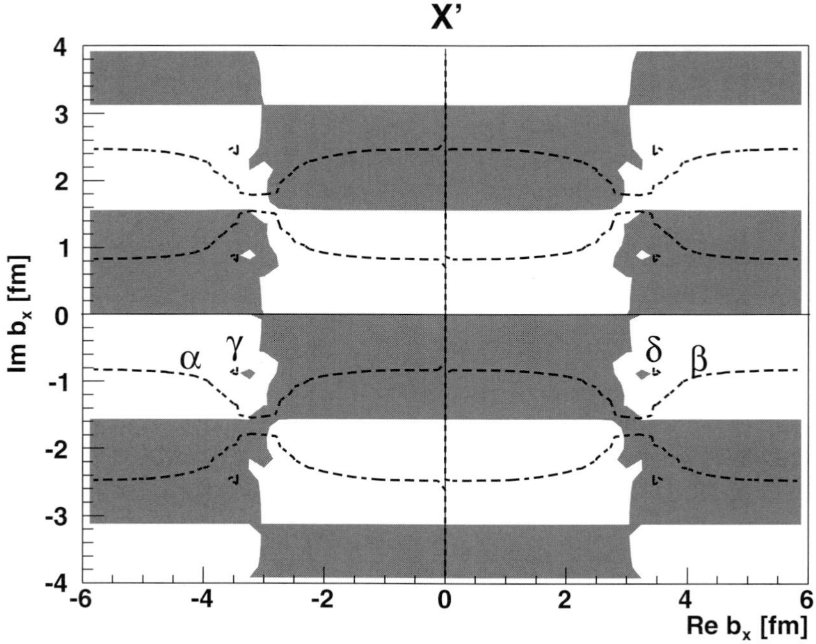

Figure 14.9 Regions of positive and negative values of $\mathrm{Re}\,X'(b_x)$, shown in the complex b_x-plane for the sum of two "Fermi" phase shift functions Eq. (14.6a), with the parameters of Eq. (14.10).

points labeled α and β, coming in from $\mathrm{Re}\,b_x = \pm\infty$, below the real axis. The sharp outer edge provides additional stationary points, labeled γ and δ in this figure. They are confined to a very small region (indicating little dependence on the momentum transfer) and are of course closer to the real axis, since the corresponding β' is smaller.

It is interesting that the trajectories extending out to $\mathrm{Re}\,b_x \to \pm\infty$, are *not* those associated with the outer structure, at $c' = 3.1$ fm $> c$, but instead those associated with the softer edge, $\beta > \beta'$. However, as $|A'_\mathrm{F}|$ increases, the trajectories associated with the singularities at $\mp c' - i\pi\beta'$ will become more important. Indeed, for $|A'_\mathrm{F}|$ of the order of 0.4 or larger (and the other parameters as defined by Eq. (14.10)), the trajectories coming in from $\mp\infty$ will end at the outermost singularities, $\mp c' - i\pi\beta'$. Then, there will, on the other hand, be trajectories of finite extensions associated with the singularities at $\mp c - i\pi\beta$.

For sufficiently large momentum transfers, when the trajectories α and β move further away from the real axis, the contributions of γ and δ will take over and dominate. This is seen in the differential cross section plot, Fig. 14.10, where we also show the reference cross section (dashed) for $A'_\mathrm{F} = 0$. The reference cross section is determined by the stationary points α and β, which are practically unchanged, in analogy with the case shown in Fig. 14.5.

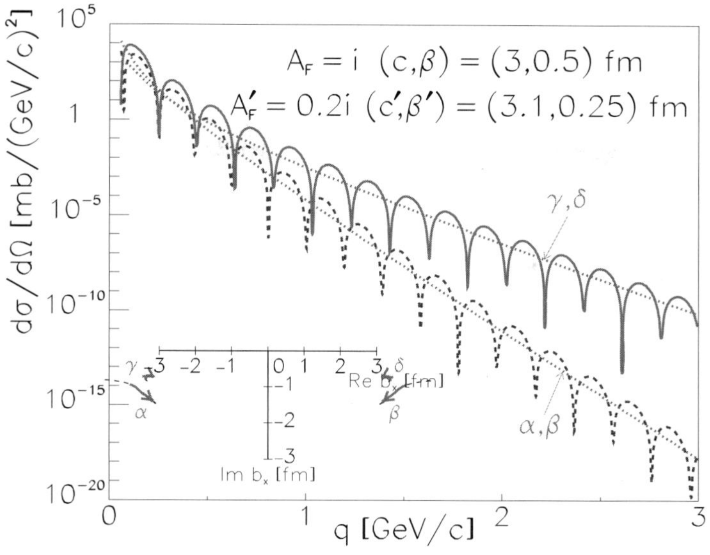

Figure 14.10 Differential cross sections corresponding to the phase shift function Eq. (14.10). The cases of $A'_F = 0.2i$ (solid) and $A'_F = 0$ (dashed) are compared. The inserts show the trajectories of stationary points for $0.02\,\text{GeV/c} \leq q \leq 5.0\,\text{GeV/c}$ (dotted). The cross sections contributed by the individual stationary points are shown as dotted curves, labeled α, β and γ, δ.

Clearly, for a smaller outer structure (a smaller value of $|A'_F|$), this "take-over" would be postponed to higher momentum transfers.

14.2.2 Sharper Internal Structure

As an example of additional, sharper internal structure, we consider the case

$$\begin{aligned} A_F &= i & c &= 3.0\text{ fm} & \beta &= 0.50\text{ fm} \\ A'_F &= 0.2i & c' &= 2.9\text{ fm} & \beta' &= 0.40\text{ fm} \end{aligned} \tag{14.11}$$

The relevant "map" of possible stationary points is shown in Fig. 14.11. As in the previous case, there are trajectories of stationary points (denoted α and β) coming in from $\text{Re}\, b_x = \mp\infty$ towards the singularities at $\mp c - i\pi\beta$, and additional stationary points γ and δ associated with the singularities at $\mp c' - i\pi\beta'$, again of finite extension.

When there are two nearby singularities, the angle of approach may deviate significantly from the more typical $\pm 45^\circ$ discussed in Chapter 8 and Eq. (14.8), since the "landscape" defined by the magnitude of the diffraction integrand

$$\exp[q\, \text{Im}\, b_x - \text{Im}\, X(b_x)] \tag{14.12}$$

can get very distorted.

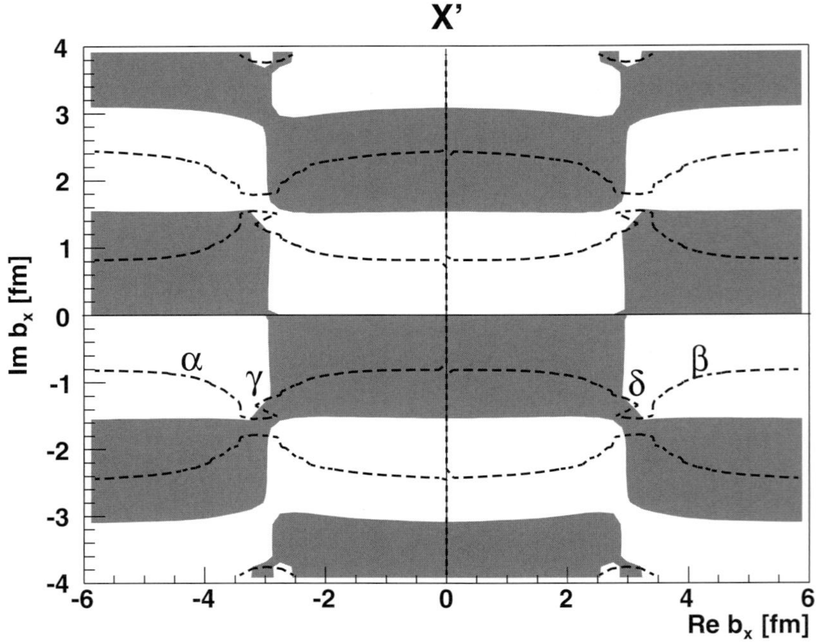

Figure 14.11 Regions of positive and negative values of Re $X'(b_x)$, shown in the complex b_x-plane for the sum of two "Fermi" phase shift functions Eq. (14.6a), with the parameters of Eq. (14.11).

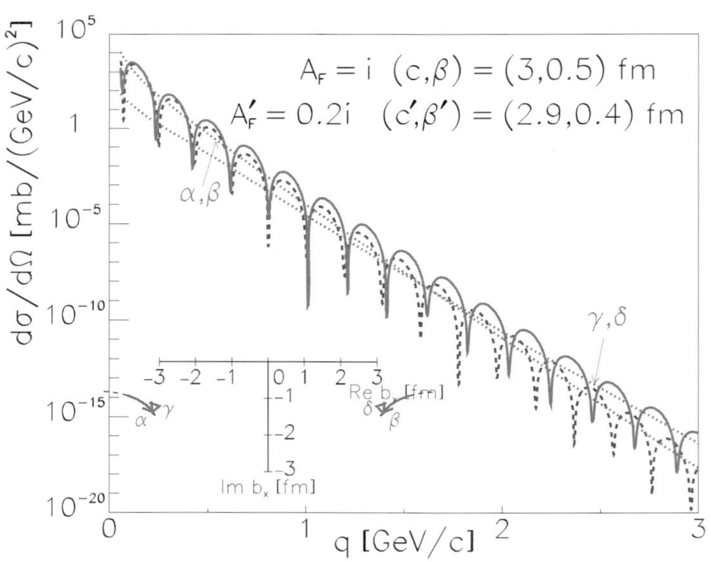

Figure 14.12 Differential cross sections corresponding to the phase shift function Eq. (14.11). The cases of $A'_{\text{F}} = 0.2i$ (solid) and $A'_{\text{F}} = 0$ (dashed) are compared. The inserts show the trajectories of stationary points for $0.02\,\text{GeV/c} \leq q \leq 5.0\,\text{GeV/c}$ (dotted). The cross sections contributed by the individual stationary points are shown as dotted curves, labeled α, β and γ, δ.

In Fig. 14.12 we show the corresponding differential cross section. It is seen that the contributions from α and β will dominate at small momentum transfers, whereas the contributions from γ and δ gradually take over at larger momentum transfers. Here, as the contributions from γ and δ start to dominate, the period of oscillation is seen to become longer, in agreement with their larger separation.

A common feature seen in many of the cases studied in this chapter is that when a "secondary" set of stationary points takes over at large momentum transfers, and perhaps provides a different period of oscillation, then this is associated with a less steep envelope of the differential cross section. The reason is that in order to become dominant, the additional structure must be sharper, meaning that the singularity must be closer to the real axis in the complex b_x-plane.

15

Different-Strength Singularities

This chapter deals with phase shift functions that have two sets of singularities, being of different degrees. The "Fermi" phase shift function

$$X(b_x) = \frac{A_{\mathrm{F}}}{1 + \exp[(b_x - c)/\beta]}, \qquad \operatorname{Re} b_x > 0, \tag{15.1}$$

will play a central role in this discussion, as a common reference. Its derivative, $X'(b_x)$ has pole singularities of second order, at $b_x = \mp c - i\pi\beta$. At large momentum transfers, the points of stationary phase will be attracted to these singularities. However, if the derivative of the overall phase shift function has a more singular point nearby, such a point may exert a stronger attraction on the stationary points. In this way, as the momentum transfer increases, the stationary points may move past those singularities associated with the ordinary "Fermi" distribution, towards other and stronger singularities. We shall in the following sections discuss a few such examples.

In this discussion, maps of the impact parameter plane are very useful for showing possible locations of points of stationary phase. In particular, the maps show where points of stationary phase are located with respect to the singularities of the phase shift function, to which they may asymptotically get attracted at large momentum transfers.

First, in Sections 15.1 and 15.2 we will consider combinations of "Fermi" functions to different (finite) powers,

$$\begin{aligned} X(b_x) &= \frac{A_{\mathrm{F}1}}{1 + \exp[(b_x - c_1)/\beta_1]} + \frac{A_{\mathrm{F}2}}{\{1 + \exp[(b_x - c_2)/\beta_2]\}^p}, \qquad \operatorname{Re} b_x > 0, \\ X(-b_x) &= X(b_x), \end{aligned} \tag{15.2}$$

with $p \neq 1$. If $p > 1$, the contribution to $X'(b_x)$ originating from the second term will be more singular, and thus exert a stronger attraction on the trajectory of stationary points. On the other hand, if $p < 1$, it will be less singular, and exert a weaker attraction. Below, we study two cases, $p = 2$ (squared "Fermi" distribution) and $p = 1/2$ (square root of "Fermi" distribution).

Then, in Section 15.3 we will consider the combination of a "Fermi" distribution, which yields a second-order singularity, together with a Gaussian one, which yields an *essential* singularity (at $(\mathrm{Re}\, b_x, \mathrm{Im}\, b_x) = (0, -i\infty)$).

We are here trying to find a function that for some range of momentum transfers leads to quasi-stable oscillations with a certain "envelope" or overall decrease of the maxima of the oscillations, but where they eventually change to a longer period and a steeper "envelope." In terms of analytic properties, we are looking for functions which have two sets of singularities. These should be located such that at moderate momentum transfers the trajectories of stationary points are attracted to the outer set (larger $|\mathrm{Re}\, b_x|$), but then at larger momentum transfers they would get more attracted to an inner set (smaller $|\mathrm{Re}\, b_x|$). For these functions to have the desired properties, the inner singularity should be stronger. Clearly, also the relative strengths A_{F1} and A_{F2} will be important.

15.1 "Fermi" and Squared "Fermi" Functions

We consider first

$$\begin{aligned} X(b_x) &= \frac{A_{\mathrm{F1}}}{1+\exp[(b_x - c_1)/\beta_1]} + \frac{A_{\mathrm{F2}}}{\{1+\exp[(b_x - c_2)/\beta_2]\}^2}, \qquad \mathrm{Re}\, b_x > 0, \\ X(-b_x) &= X(b_x), \end{aligned} \tag{15.3}$$

i.e., a "Fermi" function plus the square of another one. While the derivative of the first term has a pole singularity of second order, the second term will yield a singularity of the third order.

It is instructive to first understand scattering from the squared "Fermi" function in isolation, i.e., the case $A_{\mathrm{F1}} = 0$. A very important constraint on the location of stationary points is that $X'(b_x)$ should be real. We recall that in the neighborhood of the second-order poles of the derivative of the "Fermi" phase shift function, there are four directions in the complex b_x-plane, along which $X'(b_x)$ is real. (It is positive along every second one of these "rays".) The derivative of the squared "Fermi" function has poles of *third* order, and thus there are altogether six directions along which $X'(b_x)$ is real. It is positive along three of them.

We show in Fig. 15.1 the map of possible positions of stationary points in the complex b_x-plane, for the parameters

$$A_{\mathrm{F}} = i, \quad (c, \beta) = (3.0, 0.5)\ \mathrm{fm}. \tag{15.4}$$

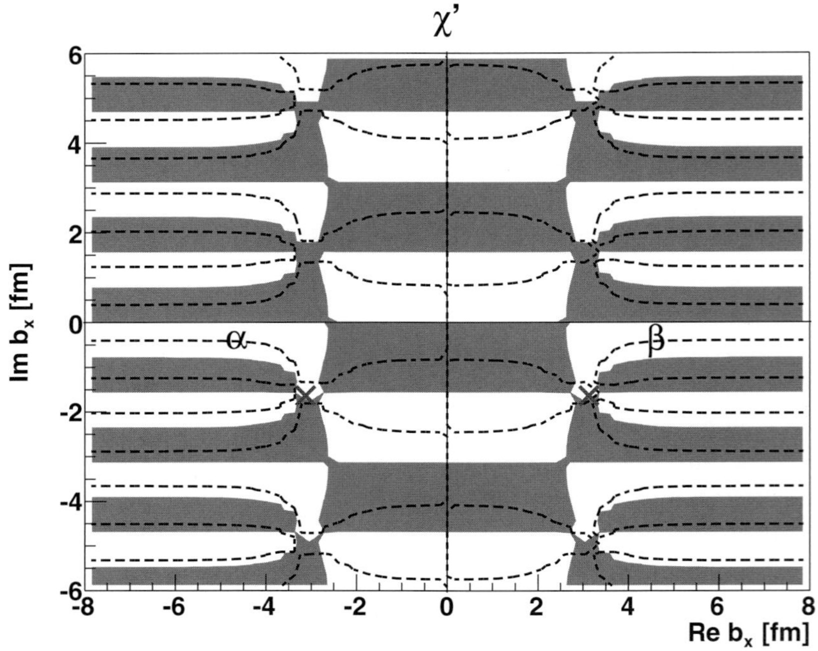

Figure 15.1 Regions of positive and negative values of Re $X'(b_x)$, shown in the complex b_x-plane for only the quadratic part of the phase shift function Eq. (15.3), i.e., with $A_{F1} = 0$. The dashed curves indicate where Im $X'(b_x) = 0$, i.e., where stationary points may be located. The crosses indicate relevant singularities of the phase shift function. The shape parameters are $(c_2, \beta_2) = (3, 0.5)$ fm.

The relevant singularities are indicated by crosses, and the segments of lines where $X'(b_x)$ is both real and positive, and also closest to the real axis, are denoted α and β.

We recall from Chapter 10 that the angle of approach to the singularity is given by the degree of the singularity, and indeed, the trajectories approach the singularities at an angle of $-\pi/2$, steeper than for the ordinary "Fermi" phase shift function, discussed in Chapter 8, where the corresponding angles of approach are $\pm\pi/4$. This means that in the "final approach," the period of oscillation will be absolutely stable, whereas there will be some steepening of the overall slope.

In Fig. 15.2 we compare the resulting differential cross section (solid) with that corresponding to the ordinary "Fermi" phase shift function (dashed). The squared "Fermi" phase shift function is seen to produce a less steeply falling cross section (with stationary points closer to the real axis), and at small momentum transfers displays some lengthening of the period of oscillation. These features are consistent with the fact that the stationary points approach the singularity at a different angle, compared to the case of the ordinary "Fermi" function. Furthermore, the less steep

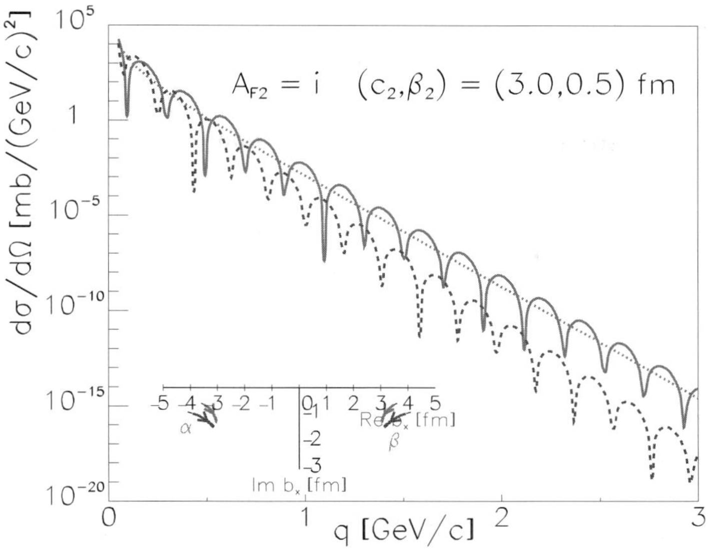

Figure 15.2 Solid: Differential cross sections corresponding to the phase shift function Eq. (15.3) with parameters corresponding to Eq. (15.4) and $A_{F1} = 0$. The inserts show the trajectories of stationary points for 0.02 GeV/c $\leq q \leq$ 5.0 GeV/c (dotted). Heavy dashed: The corresponding differential cross section for the "Fermi" phase shift function, with the same shape parameters.

envelope associated with the sharper edge reflects the fact that the effective surface thickness is $\beta_2/2$. This behavior is also reminiscent of the fact that for the limit of a black disk the differential cross section is given by the square of a Bessel function, without any exponential damping at all. Again, the asymptotic approximation agrees well with the direct numerical evaluation of the diffraction integral (not shown).

Let us now allow both terms in Eq. (15.3) to simultaneously be present. If these two singularities are suitably located with respect to each other, the stationary points may initially (for small and moderate values of momentum transfer) approach the outer and weaker ones, at $\mp c_1 - i\pi\beta_1$, but then at larger momentum transfers bypass these and move towards the other set, at $\mp c_2 - i\pi\beta_2$. However, this does not happen automatically, only for certain ranges of the parameters.

We show in Fig. 15.3 the map of possible positions of stationary points in the complex b_x-plane, for the parameters

$$A_{F1} = i, \quad (c_1, \beta_1) = (4.0, 0.5) \text{ fm},$$
$$A_{F2} = i, \quad (c_2, \beta_2) = (1.5, 1.0) \text{ fm}. \qquad (15.5)$$

The relevant singularities are indicated by crosses, and the segments of lines where $X'(b_x)$ is both real and positive, and also closest to the real axis, are denoted α, α', β, and β'.

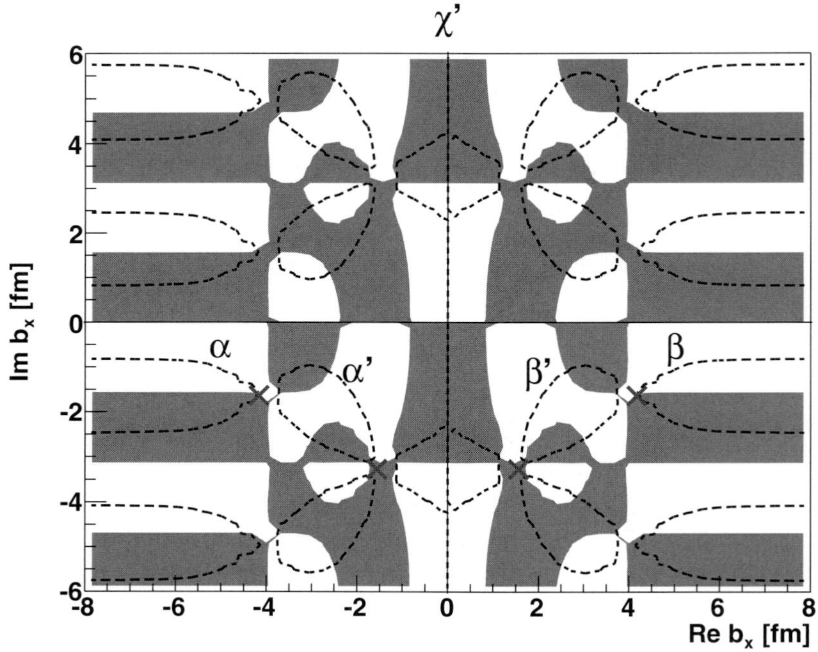

Figure 15.3 Regions of positive and negative values of $\mathrm{Re}\, X'(b_x)$, shown in the complex b_x-plane for the phase shift function Eq. (15.3). The dashed curves indicate where $\mathrm{Im}\, X'(b_x) = 0$, i.e., where stationary points may be located. The crosses indicate singularities of the phase shift function. The shape parameters are $(c_1, \beta_1) = (4, 0.5)$ fm, with $A_{\mathrm{F1}} = i$, and $(c_2, \beta_2) = (1.5, 1.0)$ fm, with $A_{\mathrm{F2}} = i$.

Away from the singularities, and the line $\mathrm{Re}\, b_x = 0$, the function $X'(b_x)$ is analytic. Hence these (dashed) lines are continuous, and do not intersect (this fact is not fully demonstrated by the available resolution).

If we gradually increase the strength associated with the more singular part of the phase shift function, i.e., make $A_{\mathrm{F2}}/A_{\mathrm{F1}}$ larger, then the shaded region (where $\mathrm{Re}\, X'(b_x) < 0$) separating them will shrink, but the two branches α and α' (as well as β and β') remain disconnected up to $A_{\mathrm{F2}}/A_{\mathrm{F1}} = \mathcal{O}(200)$. Somewhere in that range, however, they merge.

One can also make these curves move around in the complex b_x-plane by changing the shape parameters. As an illustration of this, we take the two singularities to be closer to each other, with

$$\begin{aligned} A_{\mathrm{F1}} &= i, \qquad (c_1, \beta_1) = (3, 0.5)\ \text{fm}, \\ A_{\mathrm{F2}} &= 10i, \quad (c_2, \beta_2) = (2.5, 0.75)\ \text{fm}, \end{aligned} \tag{15.6}$$

and show in Fig. 15.4 the "map" of possible trajectories of stationary points in the complex b_x-plane. The relevant singularities are again marked as crosses. The

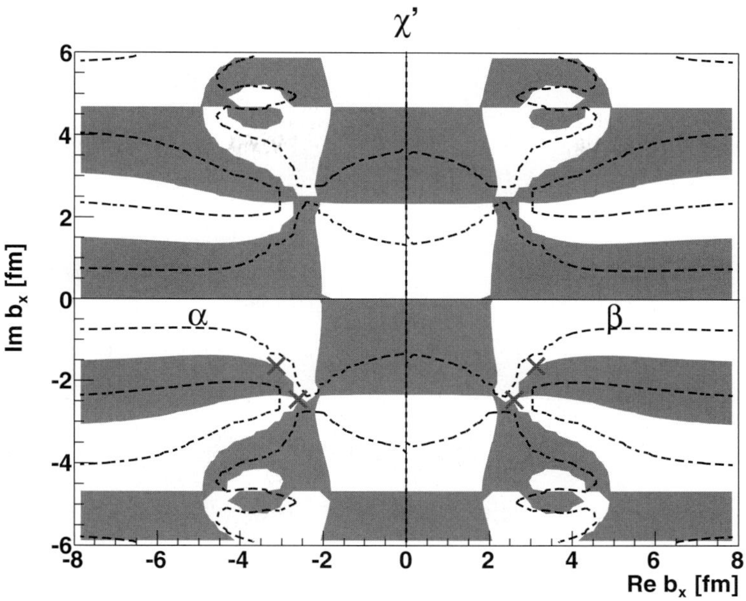

Figure 15.4 Similar to Fig. 15.3, for shape parameters $(c_1, \beta_1) = (3, 0.5)$ fm, $(c_2, \beta_2) = (2.5, 0.75)$ fm, and $A_{F2} = 10i$.

dashed lines labeled α and β are now seen to stay in the white region (where $X'(b_x) > 0$) and circumvent the outer set of singularities by passing *above*, i.e., closer to the real axis.

If we now gradually increase the value of c_2, at some point around 2.8–2.9 fm, the trajectory flips to the other side, i.e., it passes *below* the outer singularity, as illustrated in Fig. 15.5.

When it comes to constructing the differential cross section, it is important to keep in mind that this "map" is invariant if we keep the (c_i, β_i) fixed, but scale the strengths, A_{F1} and A_{F2}, as long as the ratio A_{F2}/A_{F1} is kept fixed. Such a scaling will just amount to a scaling of the momentum transfer that corresponds to a certain point on the (dashed) curve. However, while the positions of the stationary points are essential for the dependence of the differential cross section on the momentum transfer, they are not the whole story. Also the absolute value of $X(b_x)$ is relevant, and this is affected by the choices of A_{F1} and A_{F2}.

In Fig. 15.6 we show the differential cross section corresponding to the phase shift function Eq. (15.3) for the parameters adopted in Fig. 15.5, except that we have scaled down the overall strength by a factor of 10, i.e., we have taken $A_{F1} = 0.1i$ and $A_{F2} = i$. As anticipated from the discussion above, the additional and stronger singularities further away from the real axis make the envelope of the differential cross section fall in a steeper fashion.

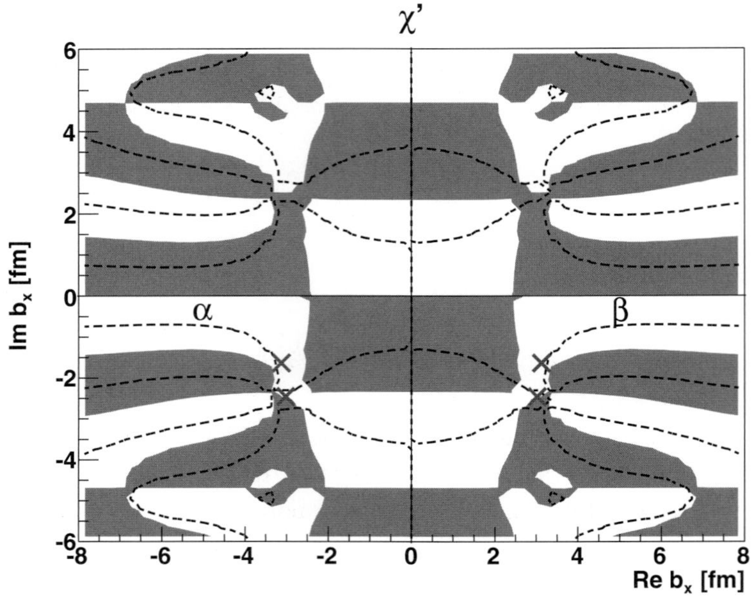

Figure 15.5 Similar to Fig. 15.4, for the phase shift function Eq. (15.3). The shape parameters are $(c_1, \beta_1) = (3, 0.5)$ fm, $(c_2, \beta_2) = (2.9, 0.75)$ fm, with $A_{F2} = 10i$.

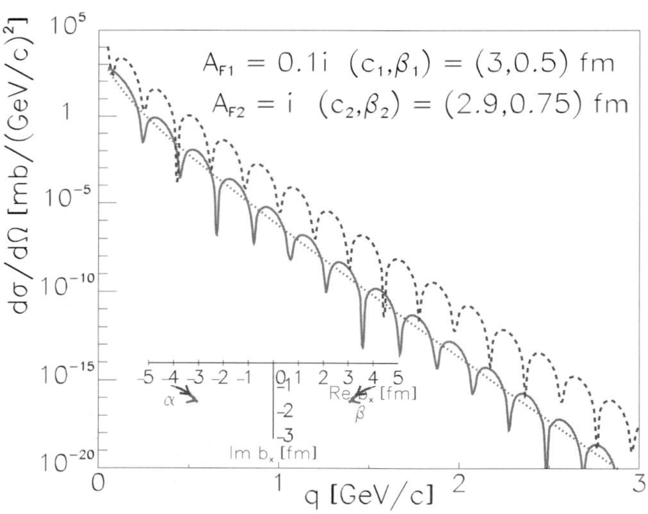

Figure 15.6 Solid: Differential cross sections corresponding to the phase shift function Eq. (15.3). The inserts show the trajectories of stationary points for $0.02\ \text{GeV/c} \le q \le 5.0\ \text{GeV/c}$ (dotted). Dashed: The corresponding differential cross section for the "Fermi" phase shift function, with parameters corresponding to Eq. (15.4).

15.2 "Fermi" and Square Root of "Fermi" Functions

We next consider another function with two sets of singularities, but where the second set of singularities now is of a lower degree compared to those of the "Fermi" function:

$$X(b_x) = \frac{A_{F1}}{1+\exp[(b_x - c_1)/\beta_1]} + \frac{A_{F2}}{\sqrt{1+\exp[(b_x - c_2)/\beta_2]}}, \qquad \mathrm{Re}\, b_x > 0,$$
$$X(-b_x) = X(b_x), \tag{15.7}$$

i.e., a "Fermi" function plus the square root of one. Of course, the square root is subject to a sign ambiguity. We choose the sign for which the square root is positive along the real b_x-axis.

Again, it is instructive to first understand scattering from the square root function, i.e., the case $A_{F1} = 0$. For the case of a pure real and repulsive (A_{F2} real and positive) square root function, the stationary points will be located on the real, negative b_x-axis. For moderate values of the momentum transfer, there will be two such points, eventually meeting at a rainbow point, beyond which one of them moves up into the second quadrant of the complex b_x-plane and becomes irrelevant, whereas the other moves down into the third quadrant. In addition, there will be a second branch (β) in the fourth quadrant, at $\mathrm{Re}\, b_x < c_2$. Below the rainbow point, it has $\mathrm{Im}\, b_x = -\pi\beta_2$.

While the square root is somewhat unfamiliar in this context, we note that the phase shift function is well-behaved at large real values of $\boldsymbol{b}$. It has in fact an exponential edge, but with a doubled surface thickness, namely $2\beta_2$.

In Fig. 15.7 we show the map of stationary points for the square root function. It has a rather unfamiliar feature, the trajectories of stationary points that emanate from the singularities at $(c_2, \beta_2) = (3, 0.5)$ fm do not continue out to $\mathrm{Re}\, b_x \to \pm\infty$. Instead, they extend *towards the central region*, namely the negative imaginary axis. This implies that the derivative $X'(b_x)$ can not become arbitrarily small along these trajectories. The smallest value, and hence the smallest value of the momentum transfer $\boldsymbol{q}$, for which we have a stationary point, is obtained for $\mathrm{Re}\, b_x \to 0$. When $\exp(-c_2/\beta_2) \ll 1$, this lowest value is given by

$$|\boldsymbol{q}|_{\min} \simeq \frac{|A_{F2}|}{2\beta_2} e^{-c_2/\beta_2}. \tag{15.8}$$

For the parameters adopted in Fig. 15.7, the value is $|\boldsymbol{q}|_{\min} \simeq 5 \times 10^{-4}$ GeV.

The trajectories of stationary points are in this case, for A_{F2} purely imaginary, seen to approach the singularities at angles of $\pm\pi/3$, as opposed to $\pm\pi/4$ in the case of the ordinary "Fermi" phase shift function.

As discussed in Chapter 8, the finite value of $X'(b_x)$ for $\mathrm{Re}\, b_x \to 0$ implies that there is an evenness issue. The cusp at $b_x \to 0$ will at some critical momentum

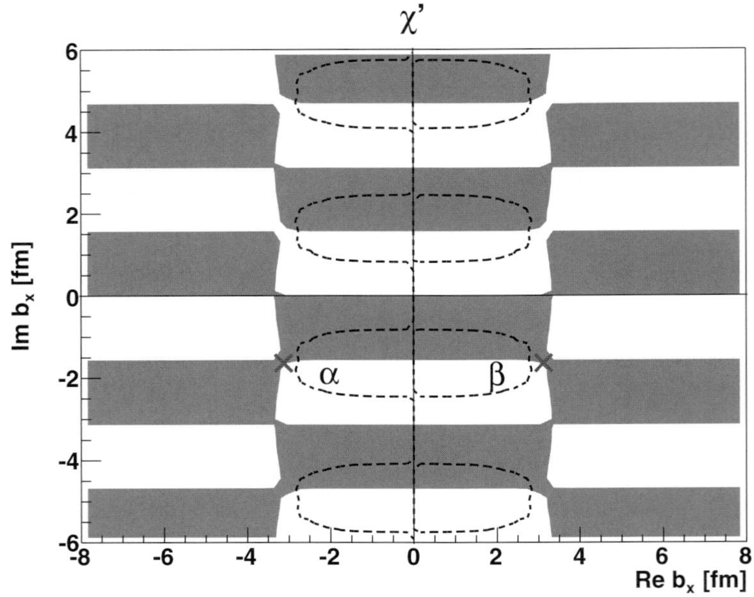

Figure 15.7 Regions of positive and negative values of $\mathrm{Re}\, X'(b_x)$, shown in the complex b_x-plane for the square root part of the phase shift function Eq. (15.7), with $A_{F1} = 0$. The dashed curves indicate where $\mathrm{Im}\, X'(b_x) = 0$, i.e., where stationary points may be located. The crosses indicate singularities of the phase shift function. The shape parameters are $(c_2, \beta_2) = (3, 0.5)$ fm.

transfer lead to a differential cross section that ceases to fall exponentially. Different ways of making $X(b_x)$ even (or flat at the origin) will lead to slightly different differential cross sections, and will not be explored here. Instead, we focus on the asymptotic formulas.

The corresponding differential cross section is shown in Fig. 15.8 for the shape parameters Eq. (15.4), and $A_{F2} = i$. Again, we compare with the ordinary "Fermi" phase shift function for the same shape parameters. For the square root function, the envelope of the differential cross section is seen to fall much more steeply, as expected because of the more diffuse surface.

Next, we consider the function Eq. (15.7) in full generality, with both A_{F1} and A_{F2} non-zero. Again, we will consider the more singular function, in this case the "Fermi" phase shift function, to have its singularities lower in the complex b_x-plane, i.e., we assume $\beta_1 > \beta_2$.

For the choice of parameters

$$A_{F1} = i, \qquad (c_1, \beta_1) = (2.7, 0.8)\ \text{fm},$$
$$A_{F2} = 0.5i, \qquad (c_2, \beta_2) = (3, 0.5)\ \text{fm}, \qquad (15.9)$$

we show in Fig. 15.9 the "map" of possible trajectories of stationary points in the complex b_x-plane. The relevant singularities are marked as crosses. The dashed

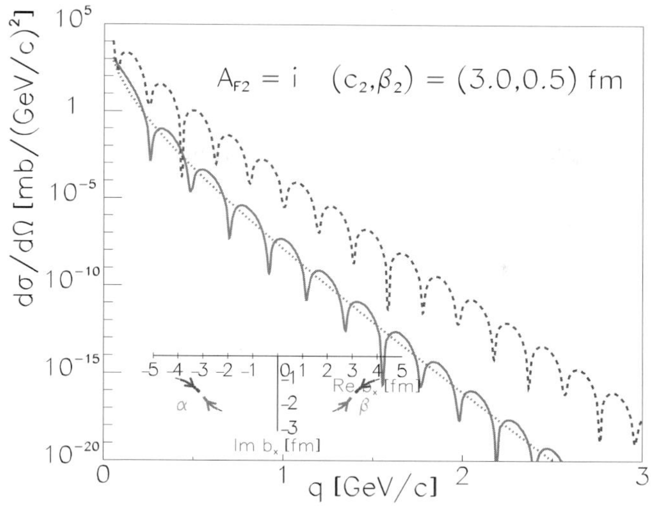

Figure 15.8 Solid: Differential cross sections corresponding to the phase shift function Eq. (15.7) for $A_{F1} = 0$, i.e., only the square root part is considered. The inserts show the trajectories of stationary points for $0.02\ \text{GeV/c} \le q \le 5.0\ \text{GeV/c}$ (dotted). Dashed: The corresponding differential cross section for the "Fermi" phase shift function, for the same parameters $(c,\beta) = (3, 0.5)$ fm.

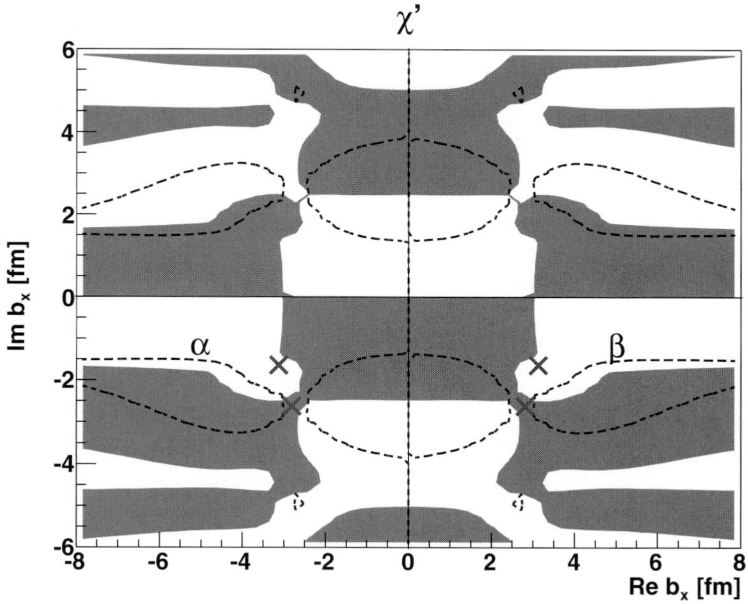

Figure 15.9 Similar to Fig. 15.7, for the phase shift function Eq. (15.7). The shape parameters are $A_{F1} = i$, $(c_1,\beta_1) = (2.7, 0.8)$ fm, and $A_{F2} = 0.5i$, $(c_2,\beta_2) = (3, 0.5)$ fm.

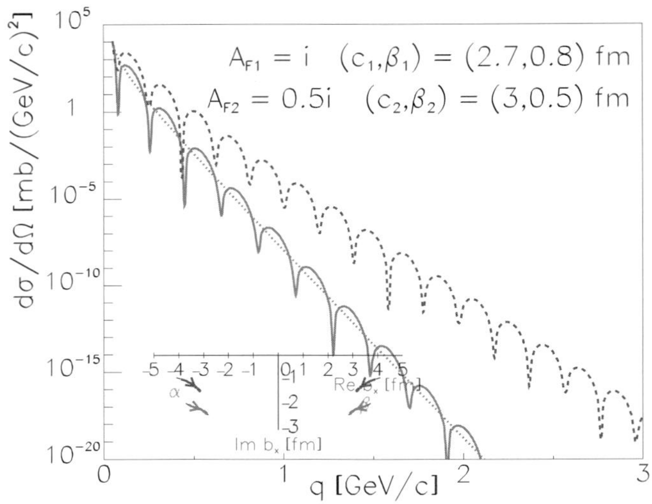

Figure 15.10 Solid: Differential cross sections corresponding to the phase shift function Eq. (15.7). The inserts show the trajectories of stationary points for $0.02\ \mathrm{GeV/c} \le q \le 5.0\ \mathrm{GeV/c}$ (dotted). Dashed: The corresponding differential cross section for the "Fermi" phase shift function, for the parameters $(c_1, \beta_1) = (3, 0.5)$ fm.

lines labeled α and β are seen to stay in the white region (where $X'(b_x) > 0$) and circumvent the outer set of singularities by passing below, i.e., further away from the real axis.

The differential cross section corresponding to the parameters of Fig. 15.9 is shown in Fig. 15.10 (solid). For comparison, we also show the differential cross section corresponding to the simple "Fermi" phase function, with the parameters of Eq. (15.4) (dashed). While the two cross section patterns are quite different, what happens here is that the second term in Eq. (15.7) has very little effect. The slope and period of the cross section are basically determined by the first term, for which $(c_1, \beta_1) = (2.7, 0.8)$ fm.

Finally, let us consider a reduced strength of the "Fermi" function relative to its square root. For the choice of parameters

$$\begin{aligned} A_{\mathrm{F1}} &= 0.5i, \quad (c_1, \beta_1) = (3, 0.75)\ \mathrm{fm}, \\ A_{\mathrm{F2}} &= i, \quad\quad (c_2, \beta_2) = (2.5, 0.5)\ \mathrm{fm}, \end{aligned} \tag{15.10}$$

we show in Fig. 15.11 the corresponding "map" of stationary points. Here, the square root part plays almost no role, even though its strength is twice that of the ordinary "Fermi" phase shift function.

In Fig. 15.12 we have reduced the strength A_{F1} even further, $A_{\mathrm{F1}} = 0.2i$. At this point, the picture is quite similar to that for the pure square root function,

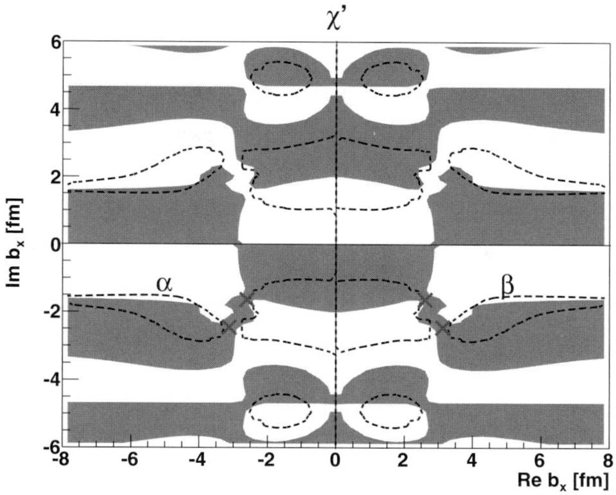

Figure 15.11 Similar to Fig. 15.9. The shape parameters are $A_{F1} = 0.5i$, $(c_1, \beta_1) = (3, 0.75)$ fm, and $A_{F2} = i$, $(c_2, \beta_2) = (2.5, 0.5)$ fm.

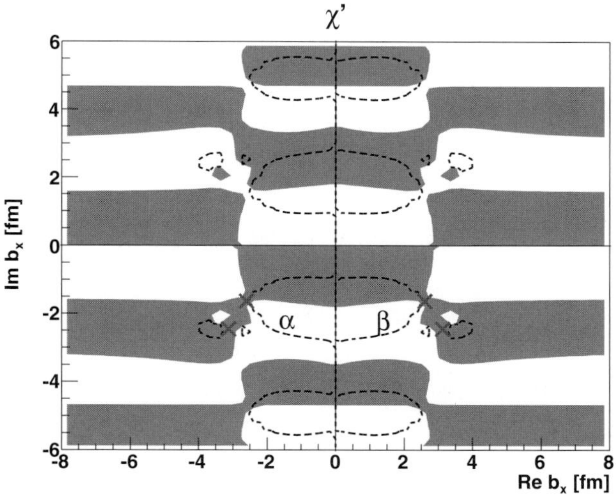

Figure 15.12 Similar to Fig. 15.11. The shape parameters are $A_{F1} = 0.2i$, $(c_1, \beta_1) = (3, 0.75)$ fm, and $A_{F2} = i$, $(c_2, \beta_2) = (2.5, 0.5)$ fm.

shown in Fig. 15.7. The ordinary "Fermi" phase shift function plays no significant role anymore.

Comparing the maps in Figs. 15.4 and 15.9 we note a qualitative difference. In Fig. 15.4 the trajectories of stationary points pass *above* (closer to the real axis) the outer singularities, whereas in Fig. 15.9 they pass *below* (further away from the real axis). This means that there must be a manifold of critical parameters. On one side of this parameter manifold the trajectories of stationary points pass above the outer singularities, on the other side they pass below.

In the terminology of Eq. (15.2), assuming we keep A_{F1}, c_1, and β_1 fixed, there is a critical value for A_{F2}, and/or c_2, and/or β_2 and/or the power p, for which the paths "jump" from one side of the singularities to the other. Such a "jump" will be accompanied by a dramatic change in the differential cross section pattern.

15.3 "Fermi" Plus Gaussian Functions

We shall next consider the composite phase shift function

$$X(b_x) = \frac{A_F}{1+\exp[(b_x - c)/\beta]} + A_G(b_x^2/\beta_G^2)^\nu \exp(-b_x^2/\beta_G^2), \qquad \mathrm{Re}\, b_x > 0,$$

$$X(-b_x) = X(b_x), \tag{15.11}$$

with $\nu = 0$, 1, or 2. We take, as a fixed reference,

$$A_F = i, \quad (c, \beta) = (3, 0.5) \text{ fm}. \tag{15.12}$$

We start the discussion by showing in Fig. 15.13 the familiar "map" of $X'(b_x)$ for the case $A_G = i$, $\nu = 0$, and $\beta_G = 1.5$ fm. For simplicity, we take both functions to be purely imaginary (on the real axis), providing stationary points symmetrically

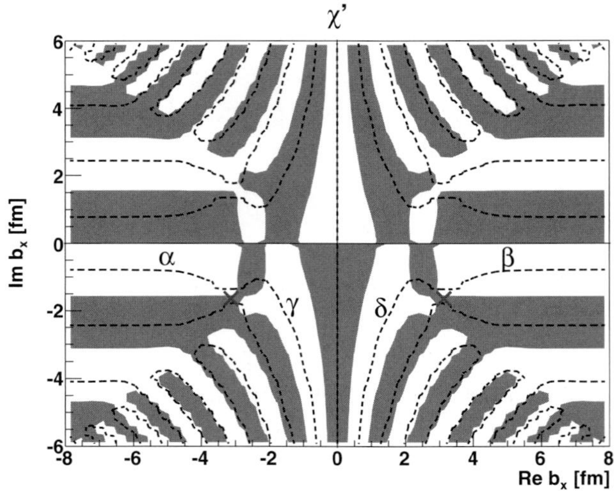

Figure 15.13 Regions of positive and negative values of $\mathrm{Re}\, X'(b_x)$, shown in the complex b_x-plane for the phase shift function Eq. (15.11). The dashed curves indicate where $\mathrm{Im}\, X'(b_x) = 0$, i.e., where stationary points may be located. The parameters of the "Fermi" function are given by Eq. (15.12), whereas those of the Gaussian are $A_G = i$, $\nu = 0$, and $\beta_G = 1.5$ fm. The crosses indicate singularities of the "Fermi" function.

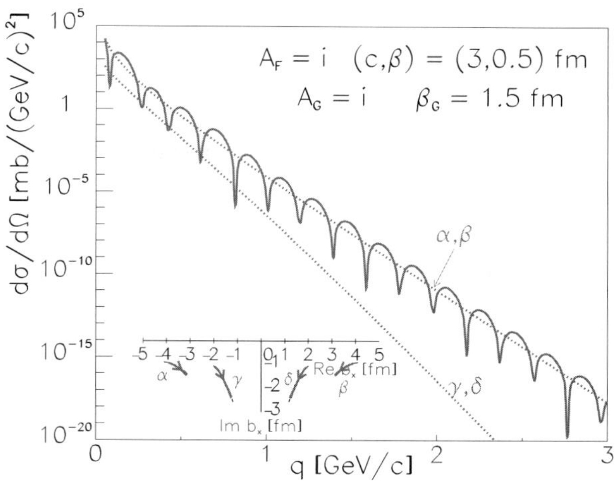

Figure 15.14 Differential cross sections corresponding to the phase shift function Eq. (15.11). The inserts show the trajectories of stationary points for $0.02\ \mathrm{GeV}/c \le q \le 5.0\ \mathrm{GeV}/c$ (dotted). The cross sections contributed by the individual stationary points are shown as dotted curves, labeled α, β and γ, δ.

located with respect to the imaginary axis. While the derivative of the "Fermi" function has second-order pole singularities in the complex plane, the Gaussian has essential singularities at $\pm i\infty$.

There are two relevant sets of stationary points. One set, (α, β), is associated primarily with the "Fermi" function, whereas the other (γ, δ) is associated primarily with the Gaussian. The essential singularity of the Gaussian will at large momentum transfers exert a strong influence on the stationary points.

The latter set is quite low in the complex plane, and thus contributes negligibly to the cross section, as shown in Fig. 15.14. It is clear that making the Gaussian more localized (smaller width β_G), the pair of stationary points (γ, δ) would be closer to the origin. This would have two effects:

i. The envelope of the amplitudes contributed by these stationary points would be less steep. At sufficiently large values of momentum transfer the scattering amplitudes contributed by these stationary points would thus become dominant.
ii. The resulting differential cross section would have a larger period of oscillations due to the smaller separation $\mathrm{Re}\,(b_{\delta x} - b_{\gamma x})$.

As mentioned above, an interesting question is whether the trajectories α and β can connect to γ and δ. In order to explore this, we let $\nu > 0$ and also take a larger value of $|A_G|$. In Figs. 15.15 and 15.16 we show the corresponding "maps" for $\nu = 1$ and $\nu = 2$. In the case of $\nu = 1$ the extra factor $(b_x/\beta_G)^2$ has the effect of pushing the stationary points γ and δ somewhat further away from the origin and the negative imaginary axis. They are however not connected to α and β.

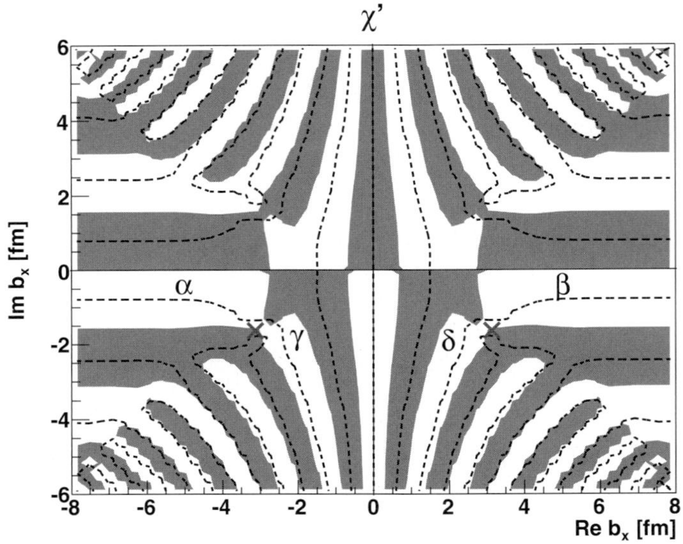

Figure 15.15 Similar to Fig. 15.13, except that for the Gaussian part $\nu = 1$.

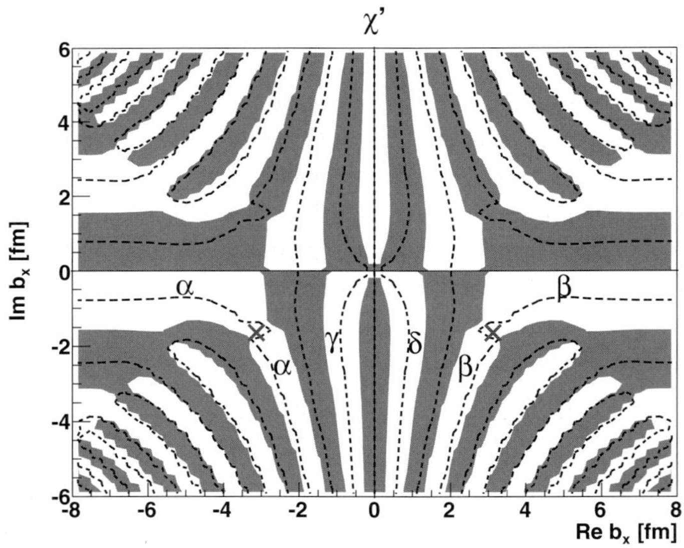

Figure 15.16 Similar to Fig. 15.13, except that for the Gaussian part $\nu = 2$.

With $\nu = 2$, on the other hand (see Fig. 15.16), the α and β trajectories continue down into the regions that were occupied by the γ and δ trajectories in the previous case, whereas a pair of "new" trajectories now appears closer to the imaginary axis. They are labeled γ and δ.

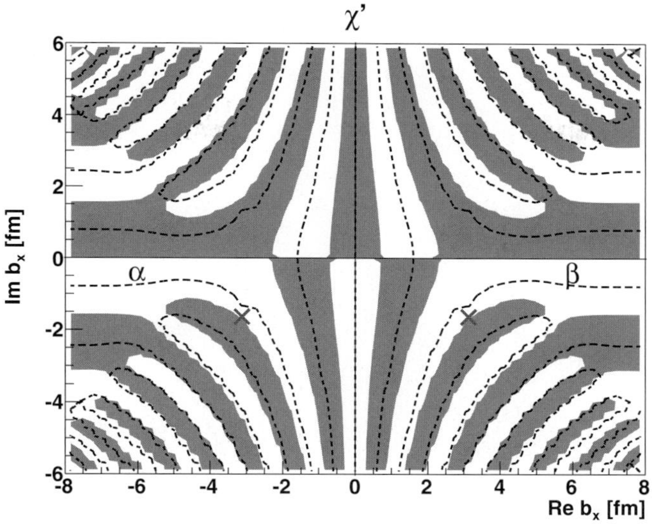

Figure 15.17 Similar to Fig. 15.15, except that $A_{\mathrm{F}} = 0.5i$ and for the Gaussian part we have $A_G = 10i$ and $\nu = 1$.

Another way to make the trajectories of stationary points α and β continue inwards towards the imaginary axis and down away from the real axis, is to increase the relative strength $|A_G|$ of the Gaussian function. This approach is illustrated in Fig. 15.17. With such a large weight on the Gaussian function, the stationary points α and β do not get "stuck" near the singularities of the "Fermi" function.

In this example, the trajectories pass on the inside of the singularities (shown as crosses in Fig. 15.17). If we instead increase the extension of the Gaussian from $\beta_G = 1.5$ fm to 3 fm, as illustrated in Fig. 15.18, the trajectories pass on the outside of the singularities, and they are also much more smooth. Of course, many intermediate situations can be imagined. If we reduce the strength of the Gaussian, stationary points will emerge near the singularities of the "Fermi" function. Eventually, as the strength of the Gaussian is further reduced, there will be two pairs of stationary points, as was the case in the initial example of Fig. 15.13 (for $A_{\mathrm{F}} = i$ and $\nu = 0$). Also, the relative magnitudes of the three shape parameters, c, β, and β_G, as well as the power ν, are important for the overall picture.

Such a smooth trajectory of stationary points going down into the complex plane, as shown in Fig. 15.18, can provide a lengthening of the period of the cross section oscillations, and a steepening of the envelope of these oscillations, as the momentum transfer increases. This is indeed the case, as shown in the corresponding differential cross section plot in Fig. 15.19. The envelope of the oscillation maxima is seen to bend down at large momentum transfers, together with some lengthening of the period.

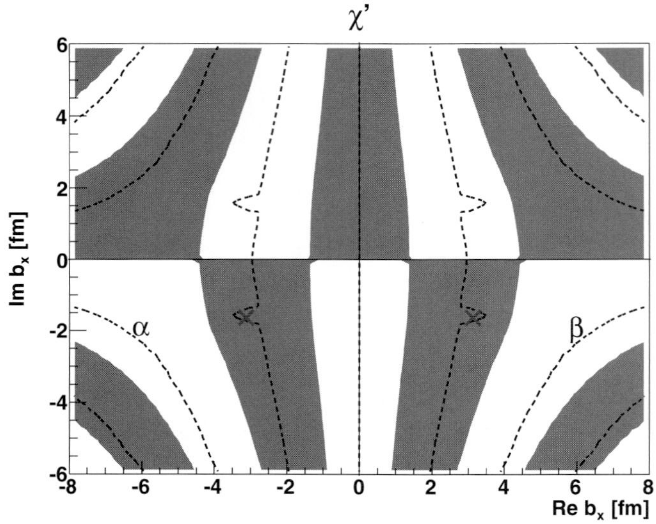

Figure 15.18 Similar to Fig. 15.17, except that for the Gaussian part we have $\beta_{\mathrm{G}} = 3.0$ fm rather than 1.5 fm.

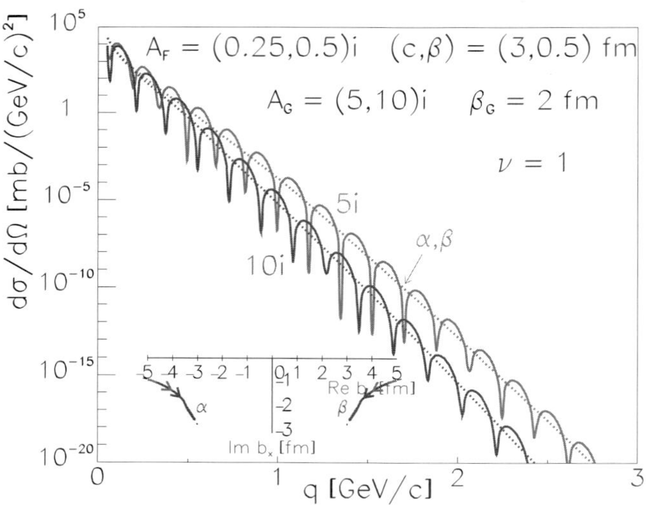

Figure 15.19 Differential cross sections corresponding to the phase shift function Eq. (15.11). The insert shows the trajectories of stationary points for $0.02\ \mathrm{GeV/c} \leq q \leq 5.0\ \mathrm{GeV/c}$ (dotted). The cross sections contributed by the individual stationary points are shown as dotted curves, labeled α, β.

In this figure we compare two cases, both having shape parameters $(c, \beta) = (3, 0.5)$ fm, $\beta_G = 2$ fm, and $\nu = 1$. They differ only in the strength parameters. The higher cross section curve refers to $A_{\mathrm{F}} = 0.25i$ and $A_G = 5i$, while the lower cross section curve refers to $A_{\mathrm{F}} = 0.5i$ and $A_G = 10i$. Since the shape parameters

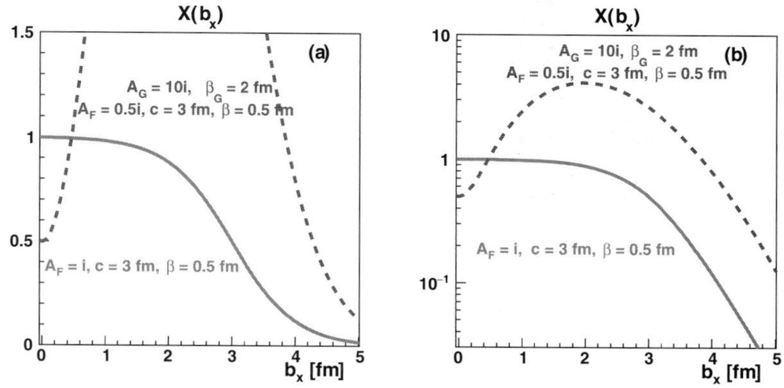

Figure 15.20 The phase shift function $X(b_x)$ of Eq. (15.11) for the parameters considered in Fig. 15.19 (dashed). For comparison, we also show the pure "Fermi" function (solid), for the parameters of Eq. 15.12. (a) linear representation; (b) logarithmic representation.

and the ratio A_F/A_G are the same in the two cases, it follows that their trajectories of stationary points are the same, as is seen from the inset. The different strength parameters lead however to different cross sections.

We have seen that for suitable parameters of the Gaussian function, the stationary points α and β continue to move inwards and down into the complex plane, without getting "stuck" at the pole singularities of the "Fermi" distribution. This leads to an interesting modification of the asymptotic behavior of the oscillations in the cross section, as compared with the stable pattern provided by the "Fermi" function.

It is a challenge to try to understand this in terms of the behavior of the phase shift function along the real axis. In Fig. 15.20 we show absolute values of phase shift functions, both in linear (a) and logarithmic (b) representations. Two cases are compared: (1) Pure "Fermi" function, i.e., $A_G = 0$, and (2) $A_G = 10i$ with $\nu = 1$ and $\beta_G = 2$ fm, and a reduced "Fermi" strength, $A_F = 0.5i$.

Attempting to understand the cross section behavior in terms of the values of the phase shift function along the real axis, two possible effects come to mind: (i) It could be due to the bulge just inside the surface region, or to (ii) a steeper behavior in the surface region. In terms of analytic functions, one can of course argue that the two aspects are related.

16

Summary

We have discussed the interpretation of hadronic scattering, at large momentum transfers, in terms of a stationary-phase approximation to the diffraction integral over impact parameters $\boldsymbol{b} = (b_x, b_y)$. For a spherically symmetric potential, the two-dimensional diffraction integral reduces to a trivial integral along the impact parameter b_y perpendicular to the scattering plane, and a nontrivial one in b_x, the component in the scattering plane. The integration over b_y essentially amounts to evaluating the integrand for $b_y = 0$, i.e., in the scattering plane.

In this limit of large momentum transfers, which admits a semiclassical approximation, the whole diffraction integral over b_x can be replaced by the sum of a few terms (typically two), corresponding to the partial amplitudes for scattering at these discrete impact parameters, determined by the stationary-phase condition

$$\boldsymbol{q} = \nabla_b \chi(\boldsymbol{b}), \tag{16.1}$$

or, with $b_y = 0$:

$$q = X'(b_x), \tag{16.2}$$

where $X(b_x) \equiv \chi(b_x, 0)$.

In the simplest possible case, that of Coulomb scattering from a point charge, the approach furnishes an exact result, up to an overall, screening-dependent and unmeasurable phase factor. For other real phase shift functions, at low momentum transfers the method typically yields two points of stationary phase. At higher momentum transfers these approach each other, and at some critical value they merge. This is rainbow scattering. Mathematically, it corresponds to $X'(b_x)$ having a maximum. In the asymptotic approach the scattering amplitude is divergent.

Beyond the rainbow (at larger momentum transfers) one stationary point has moved up into the complex plane and become irrelevant, whereas the other has moved down into the lower half-plane and yields a scattering amplitude that is exponentially damped. The reason the point of stationary phase that is in the upper complex b_x-plane is irrelevant is that the path of integration does not pass through it.

Furthermore, it is actually not difficult to find examples exhibiting more than one rainbow. Then $X'(b_x)$ must have both maxima and minima, at least one of each.

If the interaction is absorptive, then $\chi(\boldsymbol{b})$ is complex for real values of $\boldsymbol{b}$. In this case, it is clear that $\boldsymbol{b}$ (in particular, b_x) must be *complex* in order to satisfy Eqs. (16.1) and (16.2) for real $\boldsymbol{q}$.

We have illustrated characteristic features of a variety of phase shift functions having Gaussian and exponential edges, Coulomb scattering from extended charge distributions, and combinations of absorption and Coulomb scattering. Various two-scale phase shift functions have been explored, as well as phase shift functions having singularities of different degrees.

This approximation in terms of a discrete sum of terms is found to be in excellent quantitative agreement with the diffraction integral for a vast variety of phase shift functions, provided the stationary points are well separated. It offers a very physical understanding of the diffraction pattern in terms of interference between these amplitudes, often just two. For the case of pure absorption, the interpretation is particularly simple: The two points of stationary phase are symmetrically located with respect to the imaginary axis: $b_{\alpha x} = -b^*_{\beta x}$. The separation of these two stationary points, $b_{\alpha x} - b_{\beta x} = 2\,\mathrm{Re}\, b_{\alpha x}$ is then a measure of the period of oscillation, whereas $\mathrm{Im}\, b_{\alpha x}$ determines the overall slope or envelope of the oscillation pattern.

The usefulness of this observation lies in the fact that at large momentum transfers the stationary points will be located near the *singularities* of the phase shift function, as dictated by Eq. (16.1). Thus, if the diffraction pattern has a particular change or evolution with increasing momentum transfer, then this must be reflected in the movement of the points of stationary phase, attracted to one or more sets of singularities.

Appendix A

Partial Waves and Phase Shift Function

This appendix is devoted to a review of the partial wave expansion, comparing to the description of scattering in terms of a phase shift function. We start with the familiar scattering amplitude,

$$f(\theta) = \frac{1}{2ik} \sum_{\ell} (2\ell + 1)\big[e^{2i\delta_\ell} - 1\big] P_\ell(\cos\theta), \tag{A.1}$$

where $P_\ell(z)$ is the ℓ-th Legendre polynomial, and the δ_ℓ are known as phase shifts. They are often determined from the imposition of boundary conditions on certain solutions to differential equations.

For large values of ℓ and small angles, one has the asymptotic relation

$$\begin{aligned} P_\ell(\cos\theta) &= J_0\left(2(\ell + \tfrac{1}{2}) \sin \tfrac{1}{2}\theta\right) + \tfrac{1}{4} \sin^2 \tfrac{1}{2}\theta^2 + \cdots \\ &= J_0\left((\ell + \tfrac{1}{2})\theta\right) + \mathcal{O}(\theta^2) \end{aligned} \tag{A.2}$$

with $J_0(z)$ the zeroth-order Bessel function. For large values of the angular momenta, we also have the semiclassical correspondence of angular momenta,

$$kb \leftrightarrow \ell + \tfrac{1}{2}, \tag{A.3}$$

with b the impact parameter. Replacing now the sum over ℓ in Eq. (A.1) by an integral over impact parameter b, and identifying the phase shift function $\chi(b)$ as

$$\chi(b) \leftrightarrow 2\delta_\ell, \tag{A.4}$$

with b and ℓ related as in Eq. (A.3), we arrive at

$$f(\theta) = ik \int_0^\infty J_0(2kb \sin \tfrac{1}{2}\theta) \left(1 - e^{i\chi(b)}\right) b\, \mathrm{d}b. \tag{A.5}$$

This can be rewritten in several similar forms. With the momentum transfer

$$q = 2k \sin \tfrac{1}{2}\theta, \tag{A.6}$$

we find

$$f(q) = ik \int_0^\infty J_0(qb) \left(1 - e^{i\chi(b)}\right) b\, \mathrm{d}b \tag{A.7}$$

$$= \frac{ik}{2\pi} \int \mathrm{d}^2 b\, e^{-i\boldsymbol{q}\cdot\boldsymbol{b}} \left(1 - e^{i\chi(b)}\right), \tag{A.8}$$

which is the form discussed in Chapter 2. Away from the forward direction, one can drop the "1," and get

$$f(q) = \frac{k}{2\pi i} \int \mathrm{d}^2 b\, e^{-i\boldsymbol{q}\cdot\boldsymbol{b}} e^{i\chi(b)}. \tag{A.9}$$

Appendix B

Iteration for Stationary Points

This appendix gives a brief summary of how one may determine the points b_x of the stationary phase, i.e., the solutions of

$$X'(b_x) = q, \tag{B.1}$$

for $q > 0$. Explicit solutions are only available in a few special cases. Typically, one therefore has to determine the solutions by some iterative procedure. In general, the roots of Eq. (B.1) will be located off the real axis, in the complex b_x-plane. Some care therefore has to be exercised, in order to secure a convergent iteration. This may be further complicated by the fact that, as q is varied (increased, say), two solutions of Eq. (B.1) may come arbitrarily close together at a rainbow point, and then separate again (going in different directions) as q is further increased.

An easy way to survey some specific $X'(b_x)$ is to plot out the curves in the relevant part of the complex b_x-plane, where $\operatorname{Im} X'(b_x) = 0$. This can be done using some contour-plotting facility. These curves will consist of solutions to Eq. (B.1), not only for $q > 0$, but also for $q < 0$. The latter are not of interest, but the solutions for $q > 0$ and $q < 0$ may be separated if one also plots out curves where $\operatorname{Re} X'(b_x) = 0$ and determine where the two sets of curves intersect.

We are only interested in solutions to Eq. (B.1) that lie in the lower half-plane (including the real axis), $\operatorname{Im} b_x \leq 0$. Of these, usually only the two or three highest-lying ones are relevant for the actual path of integration. For any given value of q, there will thus typically be two or three solutions to Eq. (B.1) that are required. These might be distinguished by some index j, which we shall suppress in the following.

In order to establish the precise relation between b_x and q, for one particular trajectory j, we have found the following procedure useful:

(1) Consider some large, but finite value of q, which we denote $q^{(N)}$.

(2) Find some expression (even approximate) for $b_x^{(N)}$, which permits explicit evaluation, or at least iteration, for large values of $q^{(N)}$. If $X(b_x)$ has singularities, $b_x^{(N)}$ will be close to these singularities.

(3) Iterate the above expression for $b_x^{(N)}$.

(4) Improve on this solution, using, for example, the Newton–Raphson method i.e., making use of $X''(b_x)$. This is usually very efficient, and an arbitrary precision may be obtained. However, this method fails where $X'(b_x)$ is singular, which is why we start with a finite value of $q^{(N)}$.

(5) Choose a somewhat smaller q, denoted $q^{(N-1)}$. Take $b_x^{(N)}$ as a start value for the iteration, and perform steps 3 and 4.
(6) Choose a somewhat smaller q, denoted $q^{(N-2)}$. Determine the start value for iteration by extrapolation from $b_x^{(N)}$ and $b_x^{(N-1)}$. Peform step 4.

Step (6) is repeated, until one reaches a $q^{(1)}$ that is as small as the smallest value of q one is interested in. As more solutions $b_x^{(N)}, \ldots, b_x^{(N-l+1)}$ are obtained, it is convenient to determine the start point for the iteration for the next one by a second-order extrapolation from the preceding three points.

When a rainbow point is encountered, i.e., a point where $X'' = 0$, special methods are required. There are two reasons for this. Firstly, in the neighborhood of a rainbow point, $|X''|$ becomes small, and the Newton–Raphson method becomes inefficient or fails altogether. Secondly, the trajectory in the complex b_x-plane, of $b_x^{(N)}, \ldots, b_x^{(1)}$ versus $q_x^{(N)}, \ldots, q_x^{(1)}$, sharply changes direction at the rainbow point, and no reasonable start value for the iteration below the rainbow point may be obtained by a simple extrapolation from the points obtained above the rainbow point.

We here only consider rainbow points where $X''(b_x)$ has a simple zero. Because $X'(b_x)$ is analytic in some neighborhood around the rainbow point, we may write

$$X'(b_x) = X'(b_{x,\mathrm{R}}) + \tfrac{1}{2}X'''(b_{x,\mathrm{R}})(b_x - b_{x,\mathrm{R}})^2 + \cdots . \tag{B.2}$$

From this quadratic form, it follows that at the rainbow point, there are two lines along which $\operatorname{Im} X'(b_x) = 0$, intersecting at an angle of 90°. Thus, the trajectory of $b_x^{(N)}, \ldots, b_x^{(1)}$ versus $q_x^{(N)}, \ldots, q_x^{(1)}$ changes direction by $\pm 90°$ at rainbow points. To a considerable extent, this observation solves the problem of extrapolating across the rainbow point in order to determine the starting point for the iteration.

We have encountered rainbow points in the complex plane, but the more typical situation is that they occur on the real axis. This would normally happen when $X'(b_x)$ is real-valued for real b_x. With increasing q, the trajectory $b_x^{(N)}, \ldots, b_x^{(1)}$ will then at a typical rainbow point approach the real axis at a right angle for $q \gtrsim q_\mathrm{R}$, and then run along the real axis for $q < q_\mathrm{R}$. It should be noted, however, that even if $X'(b_x)$ is real for real b_x, more complicated situations may arise. In order to illustrate this, suppose that $X'(b_x)$, in addition to having maxima at $b_{x,\mathrm{R}} = b_{x,\,\mathrm{max}}$, satisfying $X'(b_{x,\,\mathrm{max}}) > 0$, also has minima $b_{x,\,\mathrm{min}}$ satisfying $X'(b_{x,\,\mathrm{min}}) > 0$. Such minima would then also be rainbow points, but with the "bright side" for $q > q_\mathrm{R}$. As q is *decreased* beyond such a rainbow point, the points of stationary phase would depart from the real b_x-axis and go off into the complex plane (at right angles with respect to the real axis).

The method may be clarified by a brief discussion of a moderately complicated example. Consider a combination of forms discussed in Chapter 12,

$$X(b_x) = \frac{\chi_0}{1 + \exp[(b_x - c)/\beta]} + 2\eta \log \sqrt{b_x^2 + a^2}, \tag{B.3}$$

which leads to

$$\begin{aligned} X'(b_x) &= \frac{-\chi_0 \exp[(b_x - c)/\beta]}{\beta\{1 + \exp[(b_x - c)/\beta]\}^2} + 2\eta \frac{b_x}{b_x^2 + a^2} \\ &\equiv X'_\mathrm{N}(b_x) + X'_\mathrm{C}(b_x). \end{aligned} \tag{B.4}$$

For some large $q^{(N)}$, the corresponding solutions to Eq. (B.1) will be close to singularities of Eq. (B.4), which are:

(i) Double poles at $b_x = \;\; c \pm (2n+1)i\pi\beta, \quad n = 0, 1, \ldots$
(ii) Double poles at $b_x = -c \pm (2n+1)i\pi\beta, \quad n = 0, 1, \ldots$
(iii) Simple poles at $b_x = \pm ia$.

There will thus, for large values of q, be two solutions in the neighborhood of each double pole, and one solution near each simple pole. Suitable expressions for the iteration discussed under points (2) and (3) above are obtained by retaining the leading terms in Eq. (B.4).

In order to determine approximately the solutions close to the double poles (i), we write Eq. (B.1) as a quadratic equation for $\exp[(b_x - c)/\beta]$,

$$\left[\exp[(b_x - c)/\beta] + 1\right]^2 + \frac{\chi_0}{\beta\left[q - X'_{\mathrm{C}}(b_x)\right]} \exp[(b_x - c)/\beta] = 0. \tag{B.5}$$

The solutions are conveniently expressed in terms of the small quantity

$$\lambda \equiv \frac{\chi_0}{2\beta\left[q - X'_{\mathrm{C}}(b_x)\right]}, \tag{B.6}$$

as

$$b_x = c - i\pi\beta + \beta \log\left[1 + \lambda + \sqrt{2\lambda + \lambda^2}\right]. \tag{B.7}$$

It should be noted that the logarithm and the square root are both multivalued, and that in Eq. (B.7) an explicit choice of branches has been made. The branch of the logarithm has been chosen such as to select the double pole that is closest to (but below) the real axis. The two branches of the square root reflect the fact that these singularities are *double* poles. There will thus be two roots of Eq. (B.5) that for $q \to \infty$ meet at these double poles; the sign in front of the square root is chosen to select the trajectory of stationary points that is closer to the real axis. The above equation is well-suited for the iteration of point (3), starting at $b_x = c - i\pi\beta$, since λ only has a weak b_x-dependence. It is not necessary to iterate Eq. (B.7) to a very high precision, since the Newton–Raphson iteration [point (4)] is more efficient. The final iteration should therefore be performed as discussed under point (4) above, i.e., using the approximation

$$b_x^{\mathrm{new}} = b_x^{\mathrm{old}} - \left[X'(b_x^{\mathrm{old}}) - q\right]/X''(b_x^{\mathrm{old}}). \tag{B.8}$$

When the phase shift function is analytic, it is less obvious where to start the iteration. If the phase shift function is "sufficiently simple," a guess based on the plot of lines where $\mathrm{Im}\, X'(b_x) = 0$ will often suffice to get started. (One has to guess not only $b_x^{(N)}$, but also $q_x^{(N)}$.) By "sufficiently simple" we mean that $X'(b_x)$ and $X''(b_x)$ should be reasonably smooth and that $|X''(b_x)|$ should not be too small in the region of complex b_x where the stationary points are searched for. If, however, $X''(b_x)$ at the guessed point is very different from its value at the stationary point, the Newton–Raphson iteration [point (4)] will usually not converge. In such cases, the following method is useful: Choose a line, starting on the real axis and going parallel to the imaginary axis out into the complex plane. This line should be chosen such that it will intersect the desired trajectory of stationary points.

Choose next a small interval, Δb_x, and investigate successive segments of the line, from $n\Delta b_x$ to $(n+1)\Delta b_x$ until a segment is found that satisfies the conditions

$$\big(\operatorname{Im} X'[n\Delta b_x]\big)\big(\operatorname{Im} X'[(n+1)\Delta b_x]\big) < 0, \tag{B.9}$$

and

$$\operatorname{Re} X'[n\Delta b_x] > 0. \tag{B.10}$$

The trajectory of stationary points will thus intersect this interval. If Δb_x is reduced, $\Delta b_x \rightarrow \Delta b_x/10$, and this procedure is repeated a couple of times, a sufficiently precise start point may be obtained as the point $\tilde{b}_x^{(N)}$ which by interpolation between $n\Delta b_x$ and $(n+1)\Delta b_x$ makes $\operatorname{Im} X'[\tilde{b}_x^{(N)}]$ vanish. The corresponding value of q is given by $\operatorname{Re} X'[\tilde{b}_x^{(N)}]$.

References

[1] Abramowitz, M., and Stegun, I. A. 1964. *Handbook of Mathematical Functions*. U.S. Government Printing Office, Washington, DC.

[2] Abramowitz, M., and Stegun, I. A. 1964. *Handbook of Mathematical Functions*. U.S. Government Printing Office, Washington, DC. Section 5.1.

[3] Bleszynski, M., Glauber, R. J., and Osland, P. 1981. Asymptotic theory of diffractive scattering. *Phys. Lett.*, **B104**, 389.

[4] Brown, B. A., and Wildenthal, B. H. 1988. Status of the nuclear shell model. *Annu. Rev. Nucl. Part. Sci.*, **38**, 29–66.

[5] Devries, H., Dejager, C. W., and Devries, C. 1974. *Nuclear Charge-Density-Distribution Parameters from Elastic Electron Scattering*. Elsevier Atomic Data and Nuclear Data Tables.

[6] Dirac, P. A. M. 1966. *Lectures on Quantum Field Theory*. Belfer Graduate School of Science, New York.

[7] Gazzaly, M. M., Hintz, N. M., Kyle, G. S., Owen, R. K., Hoffmann, G. W., et al. 1982. Neutron proton decomposition of transition matrix elements from a comparison of 800-MeV proton scattering with electromagnetic methods. *Phys. Rev.*, **C25**, 408–421.

[8] Glauber, R. J. 1959. *Lectures in Theoretical Physics*, vol. 1. Interscience Publishers, New York.

[9] Glauber, R. J. 1959. *Lectures in Theoretical Physics*, vol. 1. Interscience Publishers, New York. Page 356.

[10] Glauber, R. J. 1959. *Lectures in Theoretical Physics*, vol. 1. Interscience Publishers, New York. Page 358.

[11] Glauber, R. J. 1969. Review talk presented at the *Third International Conference on High Energy Physics and Nuclear Structure*. Columbia University, New York.

[12] Griffiths, D. J. 1998. *Introduction to Quantum Mechanics*. Prentice Hall, New Jersey.

[13] Hoffmann, G. W., Ray, L., Barlett, M., Mcgill, J., Adams, G. S., et al. 1980. 0.8-GeV p Pb-208 elastic scattering and the quantity Delta-R (NP). *Phys. Rev.*, **C21**, 1488–1494.

[14] Jackson, J. D. 1988. Simple derivation of the energy dependent effective strength of gravitational interaction in small angle scattering. *Phys. Lett.*, **B215**, 151.

[15] McVoy, K. W., and Hussein, M. S. 1984. Nearside and farside: The optics of heavy ion elastic scattering. *Pro. Part. Nucl. Phys.*, **12**, 103.

[16] Mott, N. F., and Massey, H. 1965. *The Theory of Atomic Collisions*, 3rd ed. Clarendon Press, Oxford, UK.

[17] Norton, R. E., and Abers, E. 2010. *Complex Variables for Scientists and Engineers: An Introduction*. Oxford University Press, Oxford, UK.
[18] Otsuka, T., Arima, A., and Iachello, F. 1978. Nuclear shell model and interacting bosons. *Nucl. Phys.*, **A309**, 1–33.
[19] Rutherford, E. 1911. The scattering of alpha and beta particles by matter and the structure of the atom. *Philos. Mag.*, **21**, 669–688.
[20] Shankar, R. 1994. *Principles of Quantum Mechanics, 2nd ed*. Plenum Press, New York.
[21] Sommerfeld, A. 1954. *Optics*. Academic Press, Waltham, MA.
[22] Wallace, S. J. 1975. High-energy expansion for nuclear multiple scattering. *Phys. Rev.*, **C12**, 179–193.

Index